AF389012

Plant Foods and Underutilized Fruits as Source of Functional Food Ingredients

Plant Foods and Underutilized Fruits as Source of Functional Food Ingredients: Chemical Composition, Quality Traits, and Biological Properties

Editors

Dario Donno
Federica Turrini

MDPI • Basel • Beijing • Wuhan • Barcelona • Belgrade • Manchester • Tokyo • Cluj • Tianjin

Editors
Dario Donno
Università degli Studi di Torino
Italy

Federica Turrini
Università degli Studi di Genova
Italy

Editorial Office
MDPI
St. Alban-Anlage 66
4052 Basel, Switzerland

This is a reprint of articles from the Special Issue published online in the open access journal *Foods* (ISSN 2304-8158) (available at: https://www.mdpi.com/journal/foods/special_issues/plant_foods).

For citation purposes, cite each article independently as indicated on the article page online and as indicated below:

LastName, A.A.; LastName, B.B.; LastName, C.C. Article Title. *Journal Name* **Year**, *Article Number*, Page Range.

ISBN 978-3-03943-617-0 (Hbk)
ISBN 978-3-03943-618-7 (PDF)

Cover image courtesy of Dario Donno.

Contents

About the Editors

Dario Donno, Master graduated in Chemistry and with a PhD in Agricultural, Forest and Food Sciences, is a Post-Doctoral Research Assistant at the Department of Agricultural, Forest and Food Sciences, University of Turin. His studies mainly focus on the qualitative analysis and analytical strategies (e.g., chromatographic and spectrophotometric analysis coupled to multivariate statistical analysis) for the identification and quantification of the bioactive compounds in different plant materials (buds, leaves, fruits, bark) from temperate, sub-tropical and tropical areas and agri-food industry derived-products for their quality and traceability evaluation. He is the author of more than 150 scientific and technical publications in national and international journals and books, including around 70 in Web of Science and Scopus peer-reviewed journals. He has also actively participated in national and EU research projects.

Federica Turrini is a Researcher at the Department of Pharmacy of the University of Genoa. Her research interests are in the field of: eco-compatible extractive technologies with low environmental impact and in accordance with the principles of green chemistry and green extraction; the application of environmentally friendly technologies assisted with ultrasounds and microwaves for the treatment of different food matrices and agri-food production waste; exploitation and valorization of waste from different food-processing chains in order to further formulate new potential nutraceutical and/or cosmeceutical ingredients; chemical-bromatological characterization and the assessment of the authenticity of different food matrices by untargeted spectroscopic analytical technologies coupled with chemometric techniques and targeted chromatographic characterization; formulation of new enriched and/or functional food and food dietary supplements through innovative technological treatments. In these fields of study, she is the author of many scientific publications in national and international journals and she also has participated in national and EU research projects.

Preface to "Plant Foods and Underutilized Fruits as Source of Functional Food Ingredients: Chemical Composition, Quality Traits, and Biological Properties"

It is known that specific foods provide additional health benefits to human beings as treatment and/or through the prevention of several diseases. For this reason, people have achieved a better life quality by eating fruits, vegetables, and other foods derived from plants, and using nutraceuticals, dietary supplements, or nutritional phytotherapy following official medicine. Plants produce secondary metabolites (e.g., vitamins, phenolics, terpenes, organic acids) not directly involved in normal organism development, but as defense molecules against biotic and abiotic stresses. These same compounds present specific health-promoting benefits and properties in humans and animals. Neglected and underutilized plants have become essential for the food industry, thanks to their use as an alternative for synthetic nutraceuticals and chemicals, however their potential is still not fully exploited. Studies on these raw materials are of interest to find innovative sources for natural nutraceuticals, antioxidants, and functional foods. This Special Issue provides readers with a good overview of the status and exciting developments in this field. It includes papers focused on modern analytical instrumentation and new methods and biological tests applied to the evaluation of underutilized plants and the phytochemical characterization of innovative natural sources of bioactive compounds and relative health-promoting properties. Guest Editors would like to thank all the colleagues and contributors that published their works in this Special Issue as well as the reviewers that evaluated the submissions assuring a high quality for the published studies. Guest Editors would also like to thank the publisher, MDPI, and the editorial staff of *Foods* for their high-quality, constant, and professional support as well as for their invitation to edit this Special Issue.

Dario Donno, Federica Turrini
Editors

Editorial

Plant Foods and Underutilized Fruits as Source of Functional Food Ingredients: Chemical Composition, Quality Traits, and Biological Properties

Dario Donno [1],* and Federica Turrini [2]

[1] Dipartimento di Scienze Agrarie, Forestali e Alimentari, Università degli Studi di Torino, 10095 Grugliasco (TO), Italy

[2] Dipartimento di Farmacia, Università degli Studi di Genova, 16132 Genova, Italy; turrini@difar.unige.it

* Correspondence: dario.donno@unito.it; Tel.: +39-011-670-8751

Received: 27 September 2020; Accepted: 13 October 2020; Published: 15 October 2020

Abstract: Changes in lifestyle and demographics, rising consumer incomes, and shifting preferences due to advanced knowledge about the relationships between food and health contribute to generate new needs in the food supply. Today, the role of food is not only intended as hunger satisfaction and nutrient supply but also as an opportunity to prevent nutrition-related diseases and improve physical and mental well-being. For this reason, there is a growing interest in the novel or less well-known plant foods that offer an opportunity for health maintenance. Recently, interest in plant foods and underutilized fruits is continuously growing, and agrobiodiversity exploitation offers effective and extraordinary potentialities. Plant foods could be an important source of health-promoting compounds and functional food ingredients with beneficial properties: the description of the quality and physicochemical traits, the identification and quantification of bioactive compounds, and the evaluation of their biological activities are important to assess plant food efficacy as functional foods or source of food supplement ingredients.

Keywords: natural plant foods; healthy properties; phytochemicals; agrobiodiversity; human nutrition; analytical strategies; bioactivity; unconventional fruits; in vitro test; natural antioxidants

---

It is known that specific foods confer additional health benefits to human beings as treatment and/or prevention of several diseases [1]. For this reason, people have achieved a better life quality by eating fruits, vegetables, and other foods derived from plants, and using nutraceuticals, dietary supplements, or nutritional phytotherapy following official medicine. Moreover, regional regulatory bodies stimulate high global development and research to identify new phytochemicals to be used in innovative nutraceuticals and functional foods [2]. "Nutraceuticals" may be defined as a food or part of a food that provides health-promoting benefits, and are used as adjuvants in several diseases [3]. Nutraceuticals are the fastest growing sector of the food industry with a market estimated between USD 6 billion and USD 60 billion [4] (5% growing per annum). However, there is still much confusion about nutrition in the population. In Europe and the USA, approximately 70% of people would buy specific foods to reduce the risk of diseases, but they are unable to follow dietary guidelines [5].

Nutraceuticals are detailed as food products purified, produced, or extracted from an animal or plant source (e.g., antioxidants from fish oils, blueberries, elk velvet), or produced from pressed, powdered, or dried plant material and demonstrated to present a health-promoting benefit or to protect against several chronic diseases [3,4]. Dietary supplements have more specific health roles (e.g., minerals, herbs or other botanicals, amino acids, vitamins, and other dietary ingredients) to supplement the diet by increasing the total intake of these substances [6], but they are not intended to treat disease [7]. Functional food is different from nutraceuticals and may be defined as food products used in the common diet to add beneficial health effects to the traditional nutritional ones [8]. It is important to define the key relations between the proposed concepts (nutrition, health, and technology) with the main actors involved in the processes and studies for the functional food development, namely: the nutritionist, the specialist, and the food technologist. The combination of their different skills is essential for the innovative development of these productions. These products, aimed at the maintenance of well-being, should present the highest quality standards if compared to the relative conventional products [9].

In the plant biochemistry, secondary metabolites (e.g., vitamins, polyphenols, terpenoids, organic acids), produced in the plants, are not directly involved in the normal organism growth and development, but they act as defense compounds against predators, diseases, parasites, ultraviolet radiation, and oxidants to facilitate the reproductive processes [10]. These same molecules show specific health-promoting benefits and effects in men and animals; clinical, epidemiological, in vitro, and in vivo studies have demonstrated that a diet rich in plant foods may reduce the risk of some degenerative diseases. These secondary plant metabolites with low molecular weight present excellent antioxidant and anti-inflammatory properties even if their action mechanisms vary greatly depending on the chemical structure and environment [11].

Today, nutraceuticals on the market consist of both traditional foods (e.g., vegetables, fruits, meat products, fish, grains, chocolate, and tea) and non-traditional foods (e.g., added ingredients and/or nutrients, or products derived from agricultural breeding). [4]. In recent years, wild plants (neglected and underutilized plants) have become essential for the food industry, thanks to their use as an alternative for synthetic nutraceuticals and chemicals [12], but their socio-cultural, economic, and nutritional potentials are still not fully exploited [13,14]; data on the antioxidant and health-promoting properties of several natural resources, as plants not used in medicine and nutrition, still lack. Studies on these plant materials are of interest to find innovative sources for natural nutraceuticals, antioxidants, and functional foods [15]. Recent studies in the field of natural biomolecules are increasing knowledge on naturally health-promoting substances available in food, mainly in fruit. Their use in food products may increase added-value and quality; new methodologies for extraction/purification and identification/quantification of bioactive compounds using ecofriendly analytical strategies need to be developed to improve the production yields [16,17].

This Special Issue provides readers with a good overview of the status and exciting developments in this field. It includes papers focused on modern analytical instrumentation and new methods and biological tests applied to the evaluation of underutilized plants and phytochemical characterization of innovative natural sources of bioactive compounds and relative health-promoting properties.

Fahad Alderees et al. investigated the bioactive composition of different extracts of *Tasmannia lanceolata*, *Backhousia citriodora*, and *Syzygium anisatum* by ultra-high-performance liquid chromatography and the mechanism of action against food spoilage yeasts together to their antioxidant and antimicrobial activities. The extracts showed broad-spectrum antifungal activity against weak-acid resistant yeasts in comparison to the standard antifungal agents. Polygodial, citral, and anethole were the main bioactive molecules identified in *Tasmannia lanceolata*, *Backhousia citriodora*, and *Syzygium anisatum*, respectively. The ethanol and methanol extracts showed the highest polyphenolic content and antioxidant properties, while the hexane extracts contained the highest amount of total bioactive compounds and demonstrated the strongest antimicrobial activities.

Charoonsri Chusak et al. studied the influence of the extracts from *Clitoria ternatea* L. flowers on the inhibition of pancreatic α-amylase and the starch in vitro enzymatic digestibility and predicted the glycemic index of different flours, such as potato, rice, wheat, glutinous rice, corn, and cassava flours. Moreover, the application in a bakery product, prepared from the studied flours and extracts, was also determined. The results showed that the extracts inhibited the pancreatic α-amylase activity together to a significant reduction in the glucose release, hydrolysis index (HI), and predicted glycemic index (pGI) of the considered flours.

Selina A. Fyfe et al. investigated the health-promoting properties (antioxidant capacity and antimicrobial activity), functionality, and phytochemical composition of the Australian Native Green

Plum, *Buchanania obovata* Engl., evaluating its potential as a functional ingredient in innovative food products. The seed and flesh contained several polyphenols, such as ellagic acid, p-coumaric acid, gallic acid, trans-ferulic acid, quercetin, and kaempferol, that may be responsible for the biological activities. In particular, the seed, eaten as bush food, presented a delphinidin-based anthocyanin.

Saleha Akter et al. showed the chemical and nutritional composition of the kernels of a native Australian fruit, *Terminalia ferdinandiana* (vernacular name: Kakadu Plum), as a novel nutritional source. The food industry processes the *T. ferdinandiana* fruits into puree generating seeds as a by-product that is generally discarded. This study was aimed to process the Kakadu Plum seeds separating the kernel and determining its nutritional composition. *T. ferdinandiana* kernels presented the potential to be used as a new protein source for dietary purposes and non-conventional supply of palmitic, oleic, and linoleic acids.

Yang Cao et al. presented a review on the phytochemical composition, biological properties, and nutrigenomic implications of yacon as a potential source of prebiotic, evaluating the current evidence and future directions. Yacon is an underutilized plant consumed as a traditional root-based fruit in South America and it mainly contains fructooligosaccharides (FOS) and inulin. Therefore, it has bifidogenic benefits for gut health because FOS are not easily broken down by digestive enzymes. Scientific studies on the bioactive molecules and nutrigenomic properties of the extracts and isolated compounds from yacon may help in further research to investigate yacon-based nutritional products.

Underutilized and alternative fruits represent an excellent opportunity for local growers to gain access to special or niche markets where consumers appreciate exotic traits and the presence of nutrients able to prevent degenerative diseases. The creation of specific horticultural models for fruit production may be an important opportunity to obtain a high-standardized raw material and produce high-quality fresh or derived-products. Additionally, the phytochemicals extracted from these fruits could have an excellent application in the food industry for increasing the shelf life and stability of the commercial products. Several strategies should be applied to study: (i) the toxicological traits of bioactive extracts, (ii) the metabolism of bioactive compounds (including their bioaccessibility and bioavailability), and (iii) the sensory and nutritional traits of the food products added with biologically active molecules from underutilized fruits [18]. The economic evaluation of the extraction and marketing processes should be also contemplated because these products should be environmentally safe, non-environmentally impacting, and economical [19,20].

Studies on the isolation and characterization of bioactive compounds using complementary analytical methods and on their influence on biological status in animal/human models are needed for the evaluation of their potential benefits. Finally, it is important to further confirm the lack of toxicity of these sources together to their natural bioavailability [21].

Acknowledgments: Guest Editors would like to thank all the colleagues and contributors that published their works in this Special Issue as well as the reviewers that evaluated the submissions assuring a high quality for the published studies. Guest Editors would also like to thank the publisher, MDPI, and the editorial staff of Foods for their high, constant, and professional support as well as for their invitation to edit this Special Issue.

Conflicts of Interest: The authors declare no conflict of interest.

References

1. Aluko, R.E. *Functional Foods and Nutraceuticals*; Springer: Berlin/Heidelberg, Germany, 2012.
2. Pisanello, D. *Chemistry of Foods: EU Legal and Regulatory Approaches*; Springer: Berlin/Heidelberg, Germany, 2014.
3. Srivastava, S.; Sharma, P.K.; Kumara, S. Nutraceuticals: A Review. *J. Chronother. Drug Deliv.* **2015**, *6*, 1–10.
4. Prabu, S.L.; Suriyaprakash, T.; Dinesh, K.; Suresh, K.; Ragavendran, T. Nutraceuticals: A review. *Elixir Pharm.* **2012**, *46*, 8372–8377.
5. Hardy, G. Nutraceuticals and functional foods: Introduction and meaning. *Nutrition* **2000**, *16*, 688–689. [CrossRef]

6. Roberfroid, M.B. Global view on functional foods: European perspectives. *Br. J. Nutr.* **2002**, *88*, S133–S138. [CrossRef] [PubMed]

7. Gibson, R.A.; Makrides, M. n−3 Polyunsaturated fatty acid requirements of term infants. *Am. J. Clin. Nutr.* **2000**, *71*, 251s–255s. [CrossRef] [PubMed]

8. Whitman, M.M. Understanding the perceived need for complementary and alternative nutraceuticals: Lifestyle issues. *Clin. J. Oncol. Nurs.* **2001**, *5*, 190–194. [PubMed]

9. Bigliardi, B.; Galati, F. Innovation trends in the food industry: The case of functional foods. *Trends Food Sci. Technol.* **2013**, *31*, 118–129. [CrossRef]

10. Kaur, S.; Das, M. Functional foods: An overview. *Food Sci. Biotechnol.* **2011**, *20*, 861–875. [CrossRef]

11. Komes, D.; Belščak-Cvitanović, A.; Horžić, D.; Rusak, G.; Likić, S.; Berendika, M. Phenolic composition and antioxidant properties of some traditionally used medicinal plants affected by the extraction time and hydrolysis. *Phytochem. Anal.* **2011**, *22*, 172–180. [CrossRef] [PubMed]

12. Sadia, H.; Ahmad, M.; Sultana, S.; Abdullah, A.Z.; Keat Teong, L.; Zafar, M.; Bano, A. Nutrient and mineral assessment of edible wild fig and mulberry fruits. *Fruits* **2014**, *69*, 159–166. [CrossRef]

13. Beccaro, G.L.; Bonvegna, L.; Donno, D.; Mellano, M.G.; Cerutti, A.K.; Nieddu, G.; Chessa, I.; Bounous, G. Opuntia spp. biodiversity conservation and utilization on the Cape Verde Islands. *Genet. Resour. Crop Evol.* **2015**, *62*, 21–33. [CrossRef]

14. Donno, D.; Beccaro, G.L.; Mellano, M.G.; Cerutti, A.K.; Bounous, G. Chemical fingerprint as nutraceutical quality differentiation tool in *Asimina triloba* L. fruit pulp at different ripening stages: An old species for new health needs. *J. Food Nutr. Res.* **2014**, *53*, 81–95.

15. Donno, D.; Cerutti, A.K.; Prgomet, I.; Mellano, M.G.; Beccaro, G.L. Foodomics for mulberry fruit (Morus spp.): Analytical fingerprint as antioxidants' and health properties' determination tool. *Food Res. Int.* **2015**, *69*, 179–188. [CrossRef]

16. Chemat, F.; Vian, M.A.; Cravotto, G. Green extraction of natural products: Concept and principles. *Int. J. Mol. Sci.* **2012**, *13*, 8615–8627. [CrossRef] [PubMed]

17. Oroian, M.; Escriche, I. Antioxidants: Characterization, natural sources, extraction and analysis. *Food Res. Int.* **2015**, *74*, 10–36. [CrossRef] [PubMed]

18. Ayala-Zavala, J.F.; Vega-Vega, V.; Rosas-Domínguez, C.; Palafox-Carlos, H.; Villa-Rodriguez, J.A.; Siddiqui, M.W.; Dávila-Aviña, J.E.; González-Aguilar, G.A. Agro-industrial potential of exotic fruit byproducts as a source of food additives. *Food Res. Int.* **2011**, *44*, 1866–1874. [CrossRef]

19. Barnes, S.; Prasain, J. Current progress in the use of traditional medicines and nutraceuticals. *Curr. Opin. Plant Biol.* **2005**, *8*, 324–328. [CrossRef] [PubMed]

20. Cerutti, A.K.; Beccaro, G.L.; Bruun, S.; Bosco, S.; Donno, D.; Notarnicola, B.; Bounous, G. LCA application in the fruit sector: State of the art and recommendations for environmental declarations of fruit products. *J. Clean. Prod.* **2014**, *73*, 125–135. [CrossRef]

21. Moure, A.; Cruz, J.M.; Franco, D.; Domínguez, J.M.; Sineiro, J.; Domínguez, H.; José Núñez, M.A.; Parajó, J.C. Natural antioxidants from residual sources. *Food Chem.* **2001**, *72*, 145–171. [CrossRef]

Publisher's Note: MDPI stays neutral with regard to jurisdictional claims in published maps and institutional affiliations.

 *foods*

Article

Mechanism of Action against Food Spoilage Yeasts and Bioactivity of *Tasmannia lanceolata, Backhousia citriodora* and *Syzygium anisatum* Plant Solvent Extracts

Fahad Alderees [1], Ram Mereddy [2], Dennis Webber [2], Nilesh Nirmal [1] and Yasmina Sultanbawa [1,*]

[1] Queensland Alliance for Agriculture and Food Innovation, University of Queensland, Brisbane, QLD 4108, Australia; alderees@gmail.com (F.A.); nirmalnp21@yahoo.co.in (N.N.)
[2] Department of Agriculture and Fisheries, Brisbane, QLD 4108, Australia; Ram.Mereddy@daf.qld.gov.au (R.M.); Dennis.Webber@daf.qld.gov.au (D.W.)
* Correspondence: y.sultanbawa@uq.edu.au; Tel.: +61-7-344-32471

Received: 22 September 2018; Accepted: 25 October 2018; Published: 29 October 2018

Abstract: Bioactive properties of solvent extracts of *Tasmannia lanceolata, Backhousia citriodora* and *Syzygium anisatum* investigated. The antimicrobial activities evaluated using agar disc diffusion method against two bacteria (*Escherichia coli* and *Staphylococcus aureus*) and six weak-acid resistant yeasts (*Candida albicans, Candida krusei, Dekkera anomala, Rhodotorula mucilaginosa, Saccharomyces cerevisiae* and *Schizosaccharomyces pombe*). The antioxidant activities determined using DPPH (2,2-diphenyl-1-picrylhydrazyl) free radical scavenging and reducing power assays. Quantification of major active compounds using ultra-high performance liquid chromatography. Extracts showed broad-spectrum antifungal activity against weak-acid resistant yeasts in comparison to the standard antifungal agents, fluconazole and amphotericin B. *Dekkera anomala* being the most sensitive and strongly inhibited by all extracts, while *Escherichia coli* the least sensitive. Polygodial, citral and anethole are the major bioactive compounds identified in *Tasmannia lanceolata, Backhousia citriodora* and *Syzygium anisatum*, respectively. Hexane extracts contain the highest amount of bioactive compounds and demonstrate the strongest antimicrobial activities. Methanol and ethanol extracts reveal the highest phenolic content and antioxidant properties. Fluorescence microscopic results indicate the mechanism of action of *Backhousia citriodora* against yeast is due to damage of the yeast cell membrane through penetration causing swelling and lysis leading to cell death.

Keywords: natural antimicrobial; antioxidant; mechanism of action; citral; polygodial; anethole

---

1. Introduction

The market for soft drinks and fruit juices is increasing annually with the release of new beverage products, which are gaining popularity among consumers. This market expansion has increased the challenge of addressing spoilage problems [1]. Yeasts are the most common group of microorganisms responsible for spoilage of soft drinks and fruit juices due to their ability to withstand juice acidity and resist the action of weak-acid preservatives [1,2]. Beverage industries are focusing on the application of novel antimicrobial agents derived from plant sources as an alternative solution to address beverage spoilage caused by weak-acid resistant yeasts [3]. Tasmanian pepper leaf (*Tasmannia lanceolata*), lemon myrtle (*Backhousia citriodora*) and anise myrtle (*Syzygium anisatum*) are three Australian native herbs in commercial production and there is a growing interest in their bioactive properties and assessing their potential applications as functional ingredients in the beverage industry [4].

Tasmanian pepper leaf belongs to the Winteraceae family and found in forested regions in Tasmania, Victoria [5]. It is used in food as a seasoning, flavoring, coloring and preservative and it is incorporated into personal health care products [4,6,7]. Polygodial is the major bioactive compound in Tasmanian pepper leaf responsible for its strong pungent flavor and reported to be the main contributor to the antibacterial and antifungal activities [4,5,8,9]. A chemical profiling of the essential oil of Tasmanian pepper leaf shows the following sesquiterpene compounds: polygodial (36.74%), guaiol (4.46%), calamenene (3.42%), spathulenol (1.94), drimenol (1.91%), cadina-1,4-diene (1.58%), 5-hydroxycalamenene (1.47%) bicyclogermacrene (1.15%), α-cubebene (0.88%), β-caryophyllene (0.87%), α-copaene (0.48%), cadalene (0.44%), d-cadinol (0.40%), elemol (0.39%), T-muurolol (0.39%) and germacrene-D (0.33%) [10]. Some cultivated Tasmanian pepper leaf clones are found to contain polygodial as high as 64% [10]. Phenolic compounds have been identified as the major class of antioxidant compounds in Tasmanian pepper leaf solvent and polyphenol rich extracts these include coumaric acid, cyanidin-3-glucoside, chlorogenic acid, quercetin, quercetin 3-rutinoside, cyanidin 3-rutinoside [7,11,12].

Lemon myrtle is a member of the Myrtaceae family and mainly grows in the subtropical rainforests of southeast regions of Queensland, Australia [13]. Lemon myrtle contains citral as a predominant compound with antimicrobial and insect repellent properties and is used as a cure for skin diseases [14–16]. It has a pleasant lemon flavor with mild sweet notes, it is currently one of the most cultivated and commercialized native plant species utilized in cosmetics, health products, herbal teas and flavoring agents in food and beverage systems [4,17–19]. According to reported literature, lemon myrtle essential oil contains citral (82–91%), 5-hepten-2-one,6-methyl (1.54–13.82%), 2,3-dehydro-1,8-cineole (3.52%), nerol (2.66%), germacrene B (0.2–2.18%), geraniol (0.8–1.26%), linalool (0.5–5.85%), myrcene (0.4–4.39%) and citronellal (0.25–2.19%) [15,16,20]. Phenolic compounds detected in solvent and polyphenol rich extracts of lemon myrtle contributing to antioxidant activity were myricetin, hesperetin rhamnoside hesperetin hexoside, quercetin, ellagic acid, ellagitannins and ellagic acid glycosides [11,12].

Anise myrtle also belongs to the Myrtaceae family and grows in subtropical rainforests of Bellingen and Nambucca Valleys of northeast New South Wales and some regions of Queensland [21,22]. Anise myrtle essential oil contains anethole (94.97%), methyl chavicol (4.43%), α-pinene (0.09%), 1,8-cineole (0.02%) and α-farnesene (0.07%) [23]. Anethole possesses antibacterial and antifungal activities and contributes to the intense licorice and aniseed aroma in anise myrtle leaves currently being utilized in cosmetics, savory cuisines, tea blends, body and mouth care products, alcoholic drinks and pharmaceutical industries [22,24–26]. Catechin, quercetin, hesperetin, myricetin, ellagic acid, ellagitannins and ellagic acid glycosides were the main antioxidant phenolic compounds in polyphenol rich and solvent extracts of anise myrtle leaves [11,12].

Essential oils of lemon and anise myrtle with citral and anethole as the major volatile compound have shown broad spectrum antimicrobial activity against bacteria, yeast and fungi. The minimum inhibitory concentration (MIC) ranged from 0.16 to >1.25 (% *v/v*) for lemon myrtle and 0.63 to >1.25 (% *v/v*) for anise myrtle against the following bacteria: *Staphylococcus aureus*, *Escherichia coli*, *Bacillus cereus*, *Proteus vulgaris*, *Pseudomonas aeruginosa*, *Enterobacter aerogenes*, *Acinetobacter baumannii*, *Shewanella putrefaciens* and *Listeria monocytogenes* [27]. For yeast (*Saccharomyces cerevisiae*) and fungi (*Geotricum candidum*) the MIC ranged from 0.04 to 0.08 (% *v/v*) for lemon myrtle and 0.16 to 0.08 (% *v/v*) anise myrtle respectively [27]. Hexane extracts of lemon myrtle leaves have shown anti-yeast activity against *Candida albicans*, *Candida colliculosa*, *Candida lipolytica*, *Hanseniaspora uvarum*, *Pichia anomala*, *Pichia membranifaciens*, *Rhodotorula mucilaginosa*, *Schizosaccharomyces octosporus*, however, the anise myrtle leaf hexane extracts did not show any activity against these yeasts [28]. Polygodial a major compound in Tasmanian pepper leaf extracts has shown anti yeast activity against *Zygosaccharomyces bailii* and *Saccharomyces cerevisiae*, antifungal activity against *Sclerotinia libertiana*, *Mucor mucedo*, *Rhizopus chinensis*, *Aspergillus niger*, *Penicillium crustosum* and antibacterial activity against *Salmonella choleraesuis* [6,8]. Hexane and methanol extracts of Tasmanian pepper leaf have shown anti-yeast activity against *Candida albicans*, *Candida colliculosa*, *Candida lipolytica*, *Candida stellata*, *Hanseniaspora uvarum*, *Pichia anomala*, *Pichia membranifaciens*, *Rhodotorula mucilaginosa*,

Schizosaccharomyces octosporus [28]. In the study by Zhao and Agboola [28] water extract had the least antimicrobial activity and hexane was the most potent.

Herb extracts could serve as functional ingredients in soft drinks and fruit beverages due to their antioxidant and antimicrobial properties. In recent years, much attention has been given to the application of natural compounds as an alternative solution to tackle beverage spoilage problems caused by weak-acid resistant yeasts [29–31]. The aim of this study is to measure the amounts of major compounds present in different solvent extracts of Tasmanian pepper leaf, lemon myrtle, and anise myrtle and assess the extracts bioactive properties and mechanism of action against weak-acid resistant yeasts and bacteria. This is the first study to assess the antimicrobial properties of different extracts of these herbs against a range of weak-acid resistant spoilage yeasts of importance to the beverage industry.

2. Materials and Methods

2.1. Plant Material

Lemon myrtle and anise myrtle were supplied by Australian Rainforest Products Pty Ltd. (Lismore, NSW, Australia), and Tasmanian pepper leaf supplied by Diemen Pepper (Birchs Bay, Tasmania, Australia). Herbs received as dried whole leaves and stored at $-20\,^{\circ}\text{C}$ until further use.

2.2. Milling

Leaves were separated from stems prior to the milling process. The milling was done using a Mixer Mill MM 400 (Retsch, Arzberg, Germany) which utilizes a metal ball inside a stainless-steel grinding jar that vibrates at a high horizontal speed to perform the grinding. Leaves were loaded in jars, sealed and securely positioned on the mixer instrument with a locking mechanism. Leaves were ground into a fine powder for 30 s at a vibrational frequency of 30 Hz. Milling was done on the same day of the experiment and powder used immediately for solvent extraction.

2.3. Solvent Extraction

Pressurized liquid extraction method was performed using an accelerated solvent extraction Dionex 350 instrument (Dionex, Sunnyvale, CA, USA). Twenty grams of each milled sample was mixed with 10 g of diatomaceous earth (Thermo-Fisher Scientific, Waltham, MA, USA) and placed in a 100 mL stainless steel extraction cell with a paper filter installed at the bottom end of the cell. Extraction settings were at $60\,^{\circ}\text{C}$ with five cycles under a nitrogen pressure of 1000 psi. Four extraction solvents, ethanol ($\geq$98%) hexane ($\geq$98.5%), methanol ($\geq$98%) and water were used for extraction. Each extract was collected into an amber glass bottle, filtered with No. 1 Whatman filter, transferred into a centrifuge tube and placed in a miVac DUO centrifugal vacuum concentrator (Genevac Ltd., Gardiner, NY, USA) to evaporate the organic solvent at $45\,^{\circ}\text{C}$. Dried extracts were weighed and stored at $4\,^{\circ}\text{C}$ until further use.

2.4. Microorganisms

The antimicrobial activity was assessed on two bacteria, Gram-positive *Staphylococcus aureus* (ATCC 9144) and Gram-negative *Escherichia coli* (ATCC 11775) and six yeasts. The six yeasts comprised of a standard reference strain (ISO TC 34 SC 9 Joint Working Group 5/ISO 11133) *Candida albicans* (ATCC 10231) and weak-acid preservative resistant strains which are *Candida krusei* (ATCC 6258), *Dekkera anomala* (ATCC 58985), *Rhodotorula mucilaginosa* (ATCC 20129), *Saccharomyces cerevisiae* (ATCC 38555) and *Schizosaccharomyces pombe* (ATCC 26189).

2.5. Antimicrobial

All extracted samples were screened for their antimicrobial properties using agar disc diffusion assay. Bacteria grown at $35\,^{\circ}\text{C}$ for 24 h and yeasts at $25\,^{\circ}\text{C}$ for 48 h prior to the day of the experiment.

A volume of 100 µL suspension of culture medium (10^5 cell per mL) adjusted to the appropriate density of 0.5 McFarland standard using a cuvette spectrophotometer at absorbance reading of 540 nm was inoculated on a solid media plate where standard plate count agar (Oxoid, London, UK) was used for bacteria and Sabouraud dextrose agar (Oxoid, London, UK) for yeasts. Dried sample extracts were dissolved in ethanol and 2 mg loaded onto sterile 6 mm paper discs under the fume hood and allowed to dry out before placing on the inoculated plates. Standard antibacterial chloramphenicol (Oxoid, London, UK) at 30 µg/disc and antifungal fluconazole (Sigma-Aldrich, St. Louis, MO, USA) and amphotericin B (Sigma-Aldrich, St. Louis, MO, USA) at 20 µg/disc were included as positive controls. Ethanol and water used to dissolve the standard antimicrobial drugs and assayed as negative controls. Incubation was done at 35 °C for 24 h for bacteria and at 25 °C for 48 h for yeasts. The experiments were done in duplicate. Zones of inhibition were measured using a digital caliper and expressed in millimeters. Criteria for antimicrobial strength were divided into three ranges according to Ahmad et al. [32]: weak activity (inhibition zone <10 mm), moderate activity (inhibition zone 10 to 15 mm) and strong activity (inhibition zone >15 mm).

2.6. Yeast Cell Staining and Fluorescence Microscopy

Yeast cells, *S. cerevisiae*, were grown at 25 °C for 24 h in tryptone soya yeast extract broth (Oxoid, London, UK). Yeast cells suspension was adjusted to the appropriate density of 0.5 McFarland standard using a cuvette spectrophotometer at absorbance reading of 540 nm. The adjusted suspension divided into control and treatment groups. The treatment group was centrifuged at 2500 rpm for 3 min, supernatant was removed, 4% (*v*/*v*) lemon myrtle hexane extract (dissolved in sterile water containing 0.4% tween-80) was added and allowed to stand for 30 and 60 min. The control group treated in the same manner except 0.1 M phosphate buffer added instead of lemon myrtle extract. Suspensions of treatment and control groups centrifuged at 2500 rpm for 3 min and washed with 0.1 M phosphate buffer. Cells fixed and stained according to Shimada et al. [33] with few modifications. Cells fixed for 30 min by the addition of 4% paraformaldehyde (4 mL), centrifuged (2500 rpm, 3 min) and washed twice with 0.1 M phosphate buffer. Cell suspension was mixed with 1:1 *v*/*v* of 50 ng/mL DAPI (4′,6-diamidino-2-phenylindole) (Thermo-Fisher Scientific, Waltham, MA, USA) and 8 µL of the mixture was added on a glass microscope slide and covered with a coverslip. Cell fluorescence images observed using a Leica DM6000B microscope with Leica Microsystem LAS AF6000 software (Leica, Hamburg, Germany) at 100× objective using a Leica DFC420C digital camera (Leica, Hamburg, Germany).

2.7. Total Phenolic Content

Total phenolic content (TPC) of ethanol, hexane, methanol and water extracted herb samples were spectrophotometrically analyzed according to Folin–Ciocalteu colorimetric technique [34]. Samples were diluted (10–1000 times) with Mili-Q water and 25 µL from each dilution added to the 96-well polystyrene plate (Sarstedt, Nümbrecht, Germany). Several concentrations of gallic acid (3,4,5-Trihydroxybenzoic acid ≥98%, Sigma-Aldrich, St. Louis, MO, USA), 0–100 mg/L, were prepared to construct the standard calibration curve and 25 µL of gallic acid was loaded into the plate. All wells were loaded with 125 µL of Folin-Ciocalteu's phenol reagent (Sigma-Aldrich, St. Louis, MO, USA), followed by the addition of 125 µL Sodium carbonate (Chem-Supply, Gillman, Australia). The 96-well plate placed in a microplate-reader (Tecan, Grödig, Austria) and shaken for 15 s and the absorbance reading measured at 750 nm after 15 min of incubation in the dark. Calculated results expressed as milligram of gallic acid equivalents per gram of sample dry weight (GAE/g DW).

2.8. DPPH Radical Scavenging Activity

The radical scavenging activity of the ethanol, hexane, methanol and water extracted herb samples were evaluated using DPPH (2,2-diphenyl-1-picrylhydrazyl) free radical scavenging assay according to Nirmal and Panichayupakaranant [32] with slight modifications. On the day of the experiment, DPPH (Sigma-Aldrich, St. Louis, MO, USA) concentration of 0.15 mM was prepared in 95% ethanol and mixed with different sample concentrations at 1:1 ratio (*v*/*v*) in a total volume of 3 mL. All reagents brought to room temperature before mixing. The mixing was carried out in the dark at room temperature for 30 min and the absorbance measured at 517 nm using a cuvette spectrophotometer (Thermo-Fisher Scientific, Waltham, MA, USA). Sample blank was prepared in the same manner except ethanol was used instead of DPPH solution, while control included DPPH solution and ethanol but without the addition of any sample. The percentage inhibition capacity of scavenging property for the DPPH radicals was calculated using the following formula:

$$\% \text{ inhibition} = (1 - (\text{sample absorbance}/\text{control absorbance})) \times 100. \tag{1}$$

2.9. Reducing Power

The reducing power of herb extracts analyzed as described by Nirmal and Panichayupakaranant [35]. Different concentrations of leaf extracts were prepared in a phosphate buffer (0.2 M, pH 6.6, 1 mL), mixed with 1 mL of 1% potassium ferric cyanide (Sigma-Aldrich, St. Louis, MO, USA) and allowed to incubate at 50 °C for 20 min. One milliliter of 10% trichloroacetic acid (Sigma-Aldrich, St. Louis, MO, USA) added to the mixture and centrifuged at 2500 rpm for 10 min. One milliliter from the top layer of the solution was removed and mixed with 1 ml distilled water and 0.2 mL of 0.1% ferric chloride (Sigma-Aldrich, St. Louis, MO, USA) and absorbance measured at 700 nm using a cuvette spectrophotometer. The strength of reducing power was indicated by the higher absorbance reading.

2.10. UHPLC-MS Analysis

The quantification of polygodial, citral and anethole, major compounds found in the essential oil of the tested herbs was performed by UHPLC-MS (ultra-high performance liquid chromatography/mass spectrometry) system. Extracts dissolved in methanol (HPLC Grade, Merck, Darmstadt, Germany) for analysis. The system consisted of Dionex UlitMate 3000 (Thermo-Fisher Scientific, Waltham, MA, USA) equipped with a quaternary solvent delivery system coupled to a Thermo-Fisher Q Exactive High Resolution Accurate Mass MS (Thermo-Fisher Scientific, Waltham, MA, USA). The instrument was fitted with an Acquity UHPLC BEH Shield 1.7 μm, 2.1 × 100 mm column (Waters, Milford, MA, USA). Chromatographic separation carried out with mobile phase A (Mili-Q water containing 0.1% formic acid) and mobile phase B (acetonitrile containing 0.1% formic acid). The linear gradient program was as follows: 0–0.3 min, 5% B; 0.3–2 min, 5–25% B, 2–3.5 min 25–50%, 3.5–4 min 50–80%, 4–7.5 min 80%, 7.5–8 min 80–5% and 8–11 min 5%. The flow rate was 0.5 mL/min, the column temperature was 40 °C, the autosampler temperature was 15 °C and the injection volume was 1 μL. The UHPLC-MS run in ESI-positive ion mode. Parameters were as follows: capillary temperature 268 °C; auxiliary gas heater temperature 438 °C; spray voltage 3.5 kV; sheath gas flow rate 52; auxiliary gas flow rate 14; sweep gas flow rate 3; and S-lens RF level 50. Data was collected from 3 to 10 min, over a mass range of 50–300 *m/z*, with a maximum IT of 200 ms and a resolution setting of 70,000 at 200 *m/z*. Quantification of analytes performed by external calibration, using known standard solutions of polygodial (Sigma-Aldrich, St. Louis, MO, USA), citral (Sigma-Aldrich, St. Louis, MO, USA) and anethole (Sigma-Aldrich, St. Louis, MO, USA).

2.11. Statistical Analysis

Results were expressed as mean ± standard deviation and all statistical analyses were performed using GraphPad Prism version 7.00 (GraphPad 2016, Version 7.03, GraphPad Software, Inc., San Diego, CA, USA) and figures were generated in Microsoft Excel (Office 2016, Microsoft, Redmond, WA, USA). Statistical significance of differences among treatment groups done by using one-way analysis of variance (ANOVA) followed by Tukey's test as a post hoc comparison and $p < 0.05$ is considered significant. Pearson's correlation coefficient was used to determine correlation between total phenolic content and antioxidant capacities.

3. Results

3.1. Extraction Yield and Extracts Characteristics

Yield of extractable compounds obtained from using different solvents evaluated and presented in Table 1. In general, the ranking of extraction yields obtained from different solvents are as follows in decreasing order: methanol > ethanol > water > hexane. There was a significant difference in yield between solvents used in extraction except for methanolic and ethanolic extracts of Tasmanian pepper leaf and ethanolic and water extracts of anise myrtle.

Table 1. Yields, total phenolic content, DPPH free radical scavenging and reducing power of Tasmanian pepper leaf, lemon myrtle and anise myrtle extracts.

		Methanol	Ethanol	Water	Hexane
Yields (% *w/w*)	LM	22.8 ± 0.4 [a]	17.9 ± 0.5 [b]	16.3 ± 0.6 [c]	6.41 ± 0.2 [d]
	TPL	28.3 ± 0.3 [a]	27.8 ± 0.4 [a]	25.8 ± 0.3 [b]	8.13 ± 0.5 [c]
	AM	21.8 ± 0.4 [a]	16.8 ± 0.6 [b]	17.5 ± 0.7 [b]	3.88 ± 0.3 [c]
Total phenolic content (mg GAE/gDW)	LM	419.3 ± 13.5 [a]	373.2 ± 12.6 [b]	281.7 ± 21.6 [c]	17.5 ± 1.7 [d]
	TPL	246.3 ± 17.4 [a]	215.5 ± 12.8 [a]	157.4 ± 14.6 [b]	35.7 ± 1.9 [d]
	AM	314.2 ± 17.3 [a]	310.6 ± 18.3 [a]	283.3 ± 16.5 [b]	30.5 ± 2.1 [c]
DPPH (IC_{50} µg/mL)	LM	14.4 ± 0.4 [a]	14.3 ± 0.6 [a]	31.0 ± 1.1 [b]	1678.3 ± 27.9 [c]
	TPL	36.9 ± 0.6 [a]	36.2 ± 0.8 [a]	126.4 ± 16.1 [b]	1004.7 ± 35.9 [c]
	AM	19.1 ± 1.2 [a]	21.1 ± 0.1 [a]	61.9 ± 0.2 [b]	1342.7 ± 22.9 [c]
Reducing power (Absorbance 700 nm)	0.01 mg/mL extracts — LM	0.59 ± 0.01 [a]	0.59 ± 0.02 [a]	0.32 ± 0.01 [b]	0.03 ± 0.01 [c]
	0.01 mg/mL extracts — TPL	0.29 ± 0.01 [a]	0.31 ± 0.01 [a]	0.14 ± 0.01 [b]	0.04 ± 0.02 [c]
	0.01 mg/mL extracts — AM	0.49 ± 0.02 [a]	0.45 ± 0.02 [a]	0.25 ± 0.01 [b]	0.025 ± 0.01 [c]
	0.1 mg/mL extracts — LM	1.03 ± 0.01 [a]	1.07 ± 0.02 [a]	0.56 ± 0.02 [b]	0.03 ± 0.01 [c]
	0.1 mg/mL extracts — TPL	0.51 ± 0.01 [a]	0.52 ± 0.01 [a]	0.30 ± 0.02 [b]	0.07 ± 0.01 [c]
	0.1 mg/mL extracts — AM	0.87 ± 0.1 [a]	0.84 ± 0.03 [a]	0.41 ± 0.03 [b]	0.03 ± 0.01 [c]

DPPH: 2,2-diphenyl-1-picrylhydrazyl; LM: lemon myrtle; TPL: Tasmanian pepper leaf; AM: anise myrtle; mg GAE/gDW: milligram gallic acid equivalents/g dry weight. Reducing power results expressed from testing concentrations of 0.01 and 0.1 mg/mL of extracts. Means with different letters within the same row are significantly different at $p < 0.05$.

3.2. Total Phenolic Content

The TPC of all herbs from different solvent extracts shown in Table 1. Results of the total extractable phenols expressed as mg gallic acid equivalent (GAE) per g of sample dry weight. Lemon myrtle possessed the highest TPC followed by anise myrtle and Tasmanian pepper leaf. The extraction of phenolic compounds varied among solvents used in this experiment, methanol extracts had the highest phenols followed by ethanol, water and hexane.

3.3. DPPH Radical Scavenging Activity

The free radical scavenging capacity of herb extracts from different solvents given in Table 1. The results expressed in IC_{50} (half-maximal inhibitory concentration) values are interpreted as the concentration of antioxidants required to reduce the free radical DPPH by half in the solution; lower IC_{50} values represent strong free radical scavenging activity. The strongest scavenging capacity are shown in the extracts of lemon myrtle, followed by anise myrtle and Tasmanian pepper leaf. The radical scavenging activities of extracts significantly varied with different solvents and ranked in decreasing order as methanol = ethanol > water > hexane.

3.4. Reducing Power

The reducing potential of a substance reflects its antioxidant capacity measured by utilizing the reducing power assay. The presence of antioxidant compounds (reductants) in the tested herb extracts, will react with the potassium ferricyanide (Fe^{3+}) reducing it into potassium ferrocyanide (Fe^{2+}) causing color transformation of the assay solution from yellow into different color spectrums of green and yellow which can be measured at 700 nm. Table 1 shows the reducing power of herbs extracted from different solvents. The reducing power ranking of herb extracts follows the same pattern (lemon myrtle > anise myrtle > Tasmanian pepper leaf) as the previous two assays, DPPH and TPC. The extracts reducing power potential is concentration dependent; it increases as extract concentration increases. Methanol and ethanol extracts had the highest reducing power, followed by water extracts having a moderate reducing potential while hexane extracts showed the least reducing capacity.

3.5. Relationship between Total Phenolic Content and Antioxidant Capacities

The antioxidant capacities of all tested herb extracts measured by the two assays DPPH and reducing power were found to have a significant linear correlation with the TPC. A regression analysis, correlation coefficient, performed to correlate TPC with reducing power (Figure 1a) and TPC with DPPH (Figure 1b). The overall correlation coefficient (R) between TPC and DPPH was −0.870 and between TPC and reducing power was 0.958. The correlation coefficient (R) values between TPC and DPPH were −0.988, −0.966 and −0.939 and between TPC and reducing power were 0.907, 0.979 and 0.983 for anise myrtle, lemon myrtle and Tasmanian pepper leaf, respectively. The strong relationship between TPC and antioxidant results is a clear indication that the phenolic compounds of the herb extracts contributed to the antioxidant capacity.

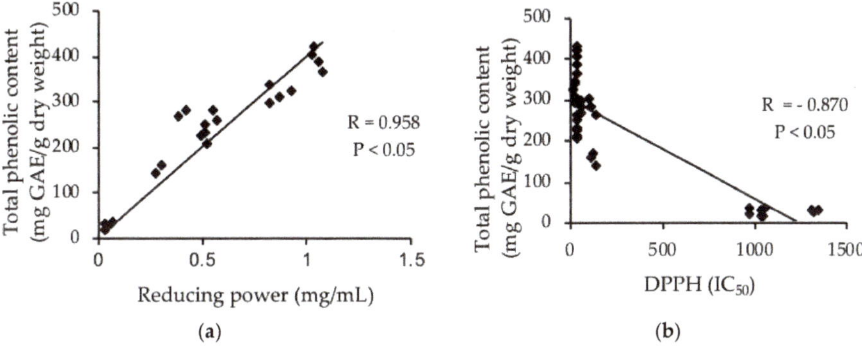

Figure 1. Correlation between total phenolic content and antioxidant capacities measured by DPPH and reducing power in Australian native herbs. (**a**) Correlation between reducing power and total phenolic content; (**b**) Correlation between DPPH and total phenolic content.

3.6. Antimicrobial Activities

The mean values and standard deviations for inhibition zones of herb extracts from various solvents and standard antimicrobial drugs against yeasts and bacteria, using agar disc diffusion method, are given in Table 2. The herb extracts from different solvents showed varied antimicrobial activities against the tested microorganisms. Hexane extracts produced the largest inhibition zones showing strong antifungal activity (inhibition zone > 15 mm) in most weak-acid resistant yeasts, followed by methanol and ethanol extracts, while water extracts showed no activity against all tested microorganisms. Extracts from hexane, methanol and ethanol showed broad-spectrum antimicrobial activity against tested microorganisms showing moderate activity (inhibition zone 10 to 15 mm) and strong activity, except for the methanolic and ethanolic extracts of lemon myrtle and anise myrtle, which did not show inhibition against *E. coli*. In addition, the hexane extract of anise myrtle showed no inhibition against *E. coli*. The standard fluconazole had broad-spectrum activity against all tested yeasts showing moderate and strong activity, while amphotericin B had a narrow-spectrum that only inhibited *D. anomala*, *S. pombe* and *R. mucilaginosa*, with inhibition zones between medium and strong activity. All hexane extracts had significantly higher zones of inhibition in comparison to fluconazole and amphotericin B at the tested concentrations, except against *D. anomala* for Tasmanian pepper leaf and anise myrtle. Furthermore, methanol and ethanol extracts of lemon myrtle and anise myrtle exhibited comparable inhibition zones to fluconazole and amphotericin B. In the case of bacteria, chloramphenicol showed significantly higher zones of inhibition than herb extracts. The negative controls, ethanol and water, produced no zones of inhibition.

Table 2. Inhibition zone (mm) of Australian native herb extracts from hexane, methanol and ethanol against yeasts and bacteria.

		D. anomala	*S. pombe*	*S. cerevisiae*	*C. albicans*	*R. mucilaginosa*	*C. krusei*	*S. aureus*	*E. coli*
TPL	M	17.6 ± 0.8 [b]	13.7 ± 0.3 [b]	17.2 ± 0.5 [b]	14.2 ± 0.3 [b]	17.1 ± 0.8 [b]	16.4 ± 0.5 [c]	12.3 ± 0.4 [b]	9.0 ± 0.2 [b]
	E	16.4 ± 0.6 [c]	13.0 ± 0.3 [b]	17.4 ± 0.4 [b]	14.8 ± 0.3 [b]	16.7 ± 0.4 [b]	15.2 ± 0.7 [b]	12.2 ± 0.2 [b]	8.3 ± 0.4 [b]
	H	23.9 ± 0.4 [a]	17.0 ± 0.3 [a]	20.7 ± 0.4 [a]	17.1 ± 0.4 [a]	21.4 ± 0.7 [a]	19.6 ± 0.5 [a]	13.6 ± 0.2 [a]	10.9 ± 0.3 [a]
LM	M	27.3 ± 0.7 [b]	12.1 ± 0.8 [b]	11.8 ± 1.2 [b]	12.7 ± 1.3 [b]	14.7 ± 0.9 [b]	11.0 ± 0.8 [b]	10.1 ± 0.4 [a,b]	0
	E	24.1 ± 0.9 [c]	11.1 ± 0.5 [b]	10.9 ± 1.0 [b]	14.1 ± 0.7 [b]	14.1 ± 0.7 [b]	10.1 ± 0.7 [b]	9.3 ± 0.7 [b]	0
	H	43.3 ± 2.1 [a]	35.7 ± 1.2 [a]	34.9 ± 1.4 [a]	26.8 ± 0.7 [a]	21.04 ± 1.8 [a]	19.8 ± 1.4 [a]	11.5 ± 0.8 [a]	8.2 ± 0.7 [a]
AM	M	23.4 ± 0.5 [b]	13.3 ± 0.7 [b]	11.3 ± 0.6 [b]	13.1 ± 0.6 [b]	14.9 ± 0.3 [b]	10.0 ± 0.3 [b]	11.4 ± 0.5 [a]	0
	E	21.9 ± 0.8 [c]	10.5 ± 0.7 [c]	9.9 ± 0.7 [c]	10.9 ± 0.9 [c]	14.2 ± 0.4 [b]	8.9 ± 0.3 [c]	8.8 ± 0.5 [b]	0
	H	26.9 ± 0.4 [a]	14.7 ± 0.4 [a]	13.2 ± 0.5 [a]	14.7 ± 0.7 [a]	16.9 ± 0.6 [a]	11.3 ± 0.8 [a]	12.5 ± 0.6 [a]	0
Fluconazole		37.2 ± 4.5	11.1 ± 0.3	11.4 ± 0.5	10.9 ± 0.9	9.5 ± 1.5	12.7 ± 0.3	NT	NT
Amphotericin B		18.4 ± 0.3	11.9 ± 0.4	0	0	12.8 ± 0.8	0	NT	NT
Chloramphenicol		NT	NT	NT	NT	NT	NT	24.0 ± 1.9	20.5 ± 0.6

TPL: Tasmanian pepper leaf; LM: lemon myrtle; AM: anise myrtle; M: methanol; E: ethanol; H: hexane; NT: not tested. Extract concentration of 2 mg/6 mm disc. Positive controls: 30 μg/6 mm disc chloramphenicol against bacteria, 20 μg/6 mm disc fluconazole and amphotericin B against yeasts. Columns sharing different letters within the same herb extract treatment are significantly different at $p < 0.05$. [a] Tasmanian pepper leaf; [b] lemon myrtle and [c] anise myrtle. Water extracts showed no inhibition. Criteria for antimicrobial activity: <10 mm, weak; 10–15 mm, moderate; >15 mm, strong.

3.7. Mode of Antifungal Action

Lemon myrtle hexane extract showed the most effective anti-yeast activity and selected to investigate the antifungal mechanism of action against *S. cerevisiae*. *S. cerevisiae* is a widely studied eukaryotic model yeast with a relatively large cell size for better observation and investigation for morphological changes [36]. The mechanism of antifungal action of hydrophobic bioactive compounds is reported as penetrating and damaging cellular cytoplasmic membrane [37,38]. Images of *S. cerevisiae* untreated, 30 min and 60 min treated cells are given in Figure 2. The untreated cells appear to have a normal elongated-oval shape with no signs of deformation or damage to cell structure (Figure 2a). Cells exposed to treatment for a period of 30 min have undergone structural modification where cells became swollen and changed from being oval into a circular shape; in addition, in some cells the membrane collapsed and reduced in size (Figure 2b). When treatment increased to 60 min, cell damage became more profound showing cells with ruptured membrane (Figure 2c). The DAPI staining observed to be brighter on damaged cells compared to non-damaged cells this could be due to the stain accumulating on the rough layer of damaged cell membrane. It was also noticeable when cells were exposed to the fluorescence microscope light source for a longer period, the stain started to fade away at a faster rate only in membrane ruptured cells. This phenomenon indicates that the stain is contained within the boundary of non-ruptured cells and is protected inside the cell and requires a longer light exposure time to fade away compared to the stain that has leaked out of the damaged cells.

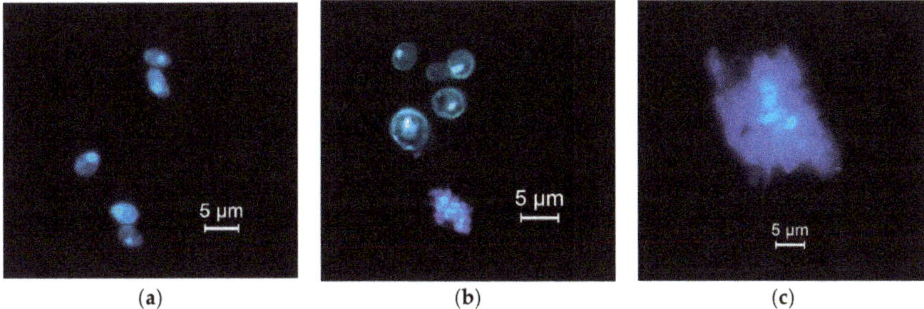

Figure 2. Illustration of morphological changes to yeast cells of *S. cerevisiae* at different treatment stages. (**a**) Untreated cells having a normal oval-shape; (**b**) Lemon myrtle extract (4% *v/v*) cell treated for 30 min showing swollen round-shaped cells next to a membrane damaged cell and (**c**) treatment for 60 min exhibits clear cell membrane rupture.

3.8. UHPLC-MS Analysis of Herb Extracts

UHPLC-MS analysis showed hexane extracts contained the highest concentration of major compounds, polygodial, citral and anethole, whereas no detectable readings were found in water extracts due to the hydrophobic characteristic of these compounds. Methanol and ethanol extracts indicate similar concentration of major compounds, which were significantly lower ($p < 0.05$) than the hexane extracts (Figure 3). The major compounds identified in herb extracts and their extracted quantities are listed as follows: polygodial found in Tasmanian pepper leaf at 41.70%, 4.43% and 3.91%; citral found in lemon myrtle at 45.7%, 9.62% and 9.29%; and anethole found in anise myrtle at 37.1%, 1.80% and 4.16%; from hexane, ethanol and methanol, respectively.

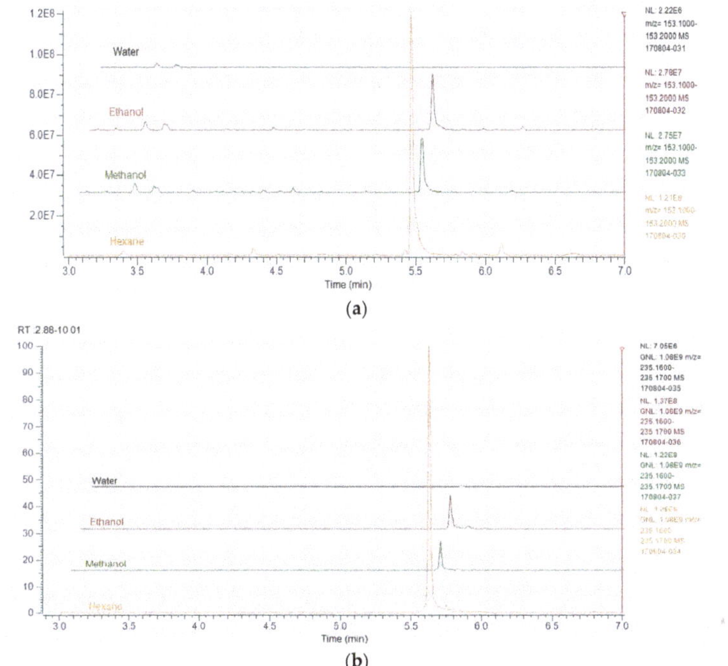

Figure 3. *Cont.*

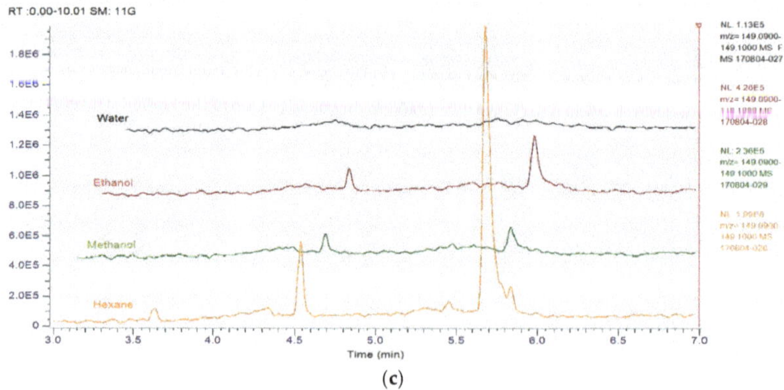

Figure 3. Chromatogram of major compounds extracted from Australian native herbs in different solvents. Citral in lemon myrtle extracts (**a**); polygodial in Tasmanian pepper leaf extracts (**b**) and anethole in anise myrtle extracts (**c**).

4. Discussion

Type of solvents used in this study had an impact on antimicrobial and antioxidant properties of herb extracts. Variation in solvent polarity is the key for the different concentrations of extracted active compounds [39]. Results showed that hexane extracts had higher antimicrobial activities compared with ethanol and methanol extracts of Tasmanian pepper leaf, lemon myrtle and anise myrtle, whereas their aqueous extracts did not exhibit antimicrobial activity. This indicates that nonpolar compounds had contributed to the antimicrobial activity of these three herbs. Previous reports found that the nonpolar compounds, polygodial, citral and anethole, were the dominant compounds in the essential oil of Tasmanian pepper leaf, lemon myrtle and anise myrtle, respectively, which could be the main contributors to the reported antibacterial and antifungal property of these herbs [10,23,24,40–42]. The antimicrobial activity of plant essential oils is often attributed to the main compounds; however, the minor compounds could contribute to antimicrobial activity and may work in synergy by forming a complex interaction with the major compounds enhancing their antimicrobial action [43,44]. For example, a study done by Sultanbawa et al. [45] evaluated the antimicrobial activity of lemon myrtle essential oil and its major bioactive compound citral against *S. aureus* and *E. coli*. The minimum inhibitory concentration of citral is 4-fold and 8-fold higher compared to the essential oil of lemon myrtle against *S. aureus* and *E. coli*, respectively.

Hexane extracts showed higher antimicrobial activity but performed poorly in the antioxidant aspect since majority of the antioxidant phenolic compounds are polar and not readily soluble in nonpolar solvents. Herb phenolic compounds have been reported to be efficiently extracted in solvents with higher polarity which makes water a superior solvent for the extraction of antioxidant compounds [46–48]. However, this was not the case in the current study since methanol and ethanol extracts contained the highest phenolic content and exerted strongest antioxidant activity, which shows the presence of some lipophilic antioxidant compounds in these herbs. Konczak et al. [12] reported that lipophilic fraction made a significant contribution to the antioxidant activity in lemon myrtle at 45.8% and anise myrtle, Tasmanian pepper leaf to a lesser degree at 5% and 14% respectively. Ellagic acid and its derivatives have been identified as the main phenolic compounds in lemon myrtle and anise myrtle extracts, while chlorogenic acid and quercetin in Tasmanian pepper leaf extract [49]. Therefore, selectivity of suitable solvents and extraction methods like ultrasonically or microwave assisted extractions for extracting phenolic compounds is important due to the diverse composition of phytochemicals in botanicals and differences in their lipophilic and hydrophilic characteristics [50,51].

The methanolic extracts of lemon myrtle, anise myrtle and Tasmanian pepper leaf showed higher TPC by 13.4, 5.6 and 2.4 fold, respectively, in comparison to the findings of Konczak et al. [12].

In addition, they reported that lemon myrtle extracts have the least antioxidant activity and phenolic content compared to anise myrtle and Tasmanian pepper leaf extracts. On the contrary, we found lemon myrtle extracts to possess the highest antioxidant activity and phenolic content. Increases in the reported TPC may be due to extraction conditions, which were done under a nitrogen pressure of 1000 psi combined with temperature of 60 °C and five extraction cycles in an accelerated solvent extraction instrument. Konczak et al. [12] sonicated herb samples for 10 min in an aqueous acidified methanol (19% water, 80% methanol, 1% hydrochloric acid) under a nitrogen atmosphere with a total of three extraction cycles. Such differences in extraction conditions between studies could have influenced the extraction and solubility of phenolic compounds [52]. Konczak et al. [12] also found a good correlation between the levels of total polyphenol content (mg GAE/g DW) and the reducing power antioxidant assays at R^2 = 0.8315 for native Australian herbs and spices which is comparable to this study. Seasonality and time of harvest are reported to significantly influence the phenolic compound content in plants, which could be another possible reason for differences between results [53].

Differences between extraction methods not only could influence the phenolic content and antioxidant activity of herbs, but also affect their antimicrobial activity. In general, direct comparison between studies of antimicrobial activity of herb extracts is challenging due to variation in methodologies including extraction conditions for bioactive compounds, and extraction concentrations. In a previous study anise myrtle extracts at a concentration of 10 mg per disc did not show any activity against *S. aureus*, *C. albicans* and *R. mucilaginosa* [28]. However, in our study anise myrtle extracts of 2 mg per disc showed moderate to strong inhibition against these microorganisms. In addition, the study found no activity against *S. aureus* and *C. albicans* from lemon myrtle methanol extract (10 mg per disc) and no activity against *E. coli* from lemon myrtle hexane extract (10 mg per disc), which is the opposite of our findings where lemon myrtle (2 mg per disc) showed antimicrobial activity against these microorganisms. In the study by Zhao and Agboola [28] the herb extractions were performed using a magnetic stirrer for 30 min at room temperature, which may not be an efficient method for the extraction of bioactive compounds. Results of extraction solvent were in agreement with Zhao and Agboola [28] in which methanol is the most efficient solvent in extracting antioxidant phenolic compounds and produced the highest extraction yield, while hexane is the weakest in both antioxidant capability and extraction yield. In addition, a similar trend in DPPH free radical scavenging capacity is supported as lemon myrtle being the strongest, followed by anise myrtle and Tasmanian pepper leaf.

Herb extracts demonstrated an overall moderate to strong antifungal activity in comparison to the standard antifungal drugs against tested yeasts. Our antibacterial results agreed with many published studies in which Gram-positive *S. aureus* is more sensitive to herb extracts than Gram-negative *E. coli* [46,54–56].

The current study showed that lemon myrtle extracts had executed their action through targeting yeast membrane in a time-dependent fashion. This observation is confirmed by Bakkali et al. [37] and Chuang et al. [57] reporting on time being required for the active compounds to partition themselves into the cell membrane and gain entry to the cell causing the observed size enlargement and swelling to the cell structure which eventually lead to rupture of cell membrane. Few studies have reported the ability of essential oils to cause swelling to bacteria, fungi and protozoa cells due to their hydrophobic properties [58–60]. Another explanation to yeast cell membrane lysis is through the binding and interacting with ergosterol found on cell membrane. Ergosterol is the main sterol component of fungal cell membrane which is responsible for the membrane rigidity, fluidity, and permeability. Therefore, binding to the ergosterol in cell membrane causes expanding of membrane lipid bilayer, altering membrane permeability and forming channels which are responsible for the cell cytoplasmic content leakage and eventual cell death [61,62]. Mechanism of action of citral, found in lemon myrtle extracts, was reported to be through the binding with ergosterol leading to membrane functional and structural destabilization [63].

5. Conclusions

In conclusion, this study reports on the antifungal activity of *Tasmannia lanceolata*, *Backhousia citriodora* and *Syzygium anisatum* extracts against weak-acid resistant yeasts and elucidates the anti-yeast mode of action of *Backhousia citriodora* extracts. The current results suggest that these three Australian native herbs possess the ability to inhibit the growth of food-spoilage yeasts that are resistant to organic weak-acids, which suggests the potential application in food and beverage industries as an alternative to synthetic antimicrobial agents.

Author Contributions: F.A.—sample preparation, antimicrobial and total phenolic testing, cell staining and fluorescence microscopic imaging, data interpretation, statistical analysis and writing of the manuscript; D.W.—UHPLC-MS analysis; N.N.—antioxidant analysis; R.M.—editing of manuscript; Y.S.—design of experiment and editing of manuscript.

Funding: Fahad Alderees's Ph.D. is supported by Public Authority for Applied Education and Training, Kuwait and Queensland Alliance for Agriculture and Food Innovation, Queensland, Australia.

Acknowledgments: The authors thank Kinnari Shelat, Andrew Cusack and Margaret Currie for their expert technical assistance.

Conflicts of Interest: The authors declare no conflict of interest. The founding sponsors had no role in the design of the study; in the collection, analyses, or interpretation of data; in the writing of the manuscript, and in the decision to publish the results.

References

1. Wareing, P. Microbiology of soft drinks and fruit juices. In *Chemistry and Technology of Soft Drinks and Fruit Juices*, 3rd ed.; Ashurst, P.R., Ed.; Wiley-Blackwell: Oxford, UK, 2016; pp. 290–309.
2. Piper, P.W. Resistance of yeasts to weak organic acid food preservatives. In *Advances in Applied Microbiology*; Laskin, A.I., Sariaslani, S., Gadd, G.M., Eds.; Elsevier: Amsterdam, The Netherlands, 2011; Volume 77, pp. 97–113.
3. Hayashi, M.; Naknukool, S.; Hayakawa, S.; Ogawa, M.; Ni'matulah, A.-B.A. Enhancement of antimicrobial activity of a lactoperoxidase system by carrot extract and β-carotene. *Food Chem.* **2012**, *130*, 541–546. [CrossRef]
4. Clarke, M. *Australian Native Food Industry Stocktake*; RIRDC Publication No. 12/066; Union Offset Printing: Canberra, Australia, 2012.
5. Pengelly, A. Indigenous and naturalised herbs: *Tasmannia lanceolata*: Mountain pepper. *Aust. J. Med. Herbal.* **2002**, *14*, 71–74.
6. Sultanbawa, Y. Tasmanian Pepper Leaf (*Tasmannia lanceolata*) Oils. In *Essential Oils in Food Preservation, Flavor and Safety*; Preedy, V.R., Ed.; Elsevier: Amsterdam, The Netherlands, 2016; pp. 819–823.
7. Cock, I.E. The phytochemistry and chemotherapeutic potential of *Tasmannia lanceolata* (*Tasmanian pepper*): A review. *Pharmacogn. Commun.* **2013**, *3*, 1–13. [CrossRef]
8. Fujita, K.-I.; Kubo, I. Naturally occurring antifungal agents against Zygosaccharomyces bailii and their synergism. *J. Agric. Food Chem.* **2005**, *53*, 5187–5191. [CrossRef] [PubMed]
9. Kubo, I.; Fujita, K.I.; Lee, S.H.; Ha, T.J. Antibacterial activity of polygodial. *Phytother. Res.* **2005**, *19*, 1013–1017. [CrossRef] [PubMed]
10. Menary, R.C.; Dragar, V.A.; Thomas, S.; Read, C.D. *Mountain Pepper Extract, Tasmannia Lanceolata: Quality Stabilisation and Registration: A Report for the Rural Industries Research and Development Corporation*; RIRDC: Barton, Australia, 2003.
11. Guo, Y.; Sakulnarmrat, K.; Konczak, I. Anti-inflammatory potential of native Australian herbs polyphenols. *Toxicol. Rep.* **2014**, *1*, 385–390. [CrossRef] [PubMed]
12. Konczak, I.; Zabaras, D.; Dunstan, M.; Aguas, P. Antioxidant capacity and phenolic compounds in commercially grown native Australian herbs and spices. *Food Chem.* **2010**, *122*, 260–266. [CrossRef]
13. Buchaillot, A.; Caffin, N.; Bhandari, B. Drying of lemon myrtle (*Backhousia citriodora*) leaves: Retention of volatiles and color. *Dry. Technol.* **2009**, *27*, 445–450. [CrossRef]
14. Horn, T.; Barth, A.; Rühle, M.; Häser, A.; Jürges, G.; Nick, P. Molecular diagnostics of Lemon Myrtle (*Backhousia citriodora* versus *Leptospermum citratum*). *Eur. Food Res. Technol.* **2012**, *234*, 853–861. [CrossRef]

15. Forbes-Smith, M.; Paton, J. *Innovative Products from Australian Native Foods*; Rural Industries Research and Development Corporation (RIRDC): Canberra, Australia, 2002.

16. Southwell, I.A.; Russell, M.; Smith, R.L.; Archer, D.W. *Backhousia citriodora* F. Muell. (*Myrtaceae*), a superior source of citral. *J. Essent. Oil Res.* **2000**, *12*, 735–741. [CrossRef]

17. Scheman, A.; Scheman, N.; Rakowski, E.-M. European directive fragrances in natural products. *Dermatitis* **2014**, *25*, 51–55. [CrossRef] [PubMed]

18. Ress, N.; Hailey, J.; Maronpot, R.; Bucher, J.; Travlos, G.; Haseman, J.; Orzech, D.; Johnson, J.; Hejtmancik, M. Toxicology and carcinogenesis studies of microencapsulated citral in rats and mice. *Toxicol. Sci.* **2003**, *71*, 198–206. [CrossRef] [PubMed]

19. Smyth, H.E.; Sanderson, J.E.; Sultanbawa, Y. Lexicon for the Sensory Description of Australian Native Plant Foods and Ingredients. *J. Sens. Stud.* **2012**, *27*, 471–481. [CrossRef]

20. Pengelly, A. Antimicrobial activity of lemon myrtle and tea tree oils. *Aust. J. Med. Herbal.* **2003**, *15*, 9.

21. Southwell, I.; Russell, M.; Smith, R. Chemical composition of some novel aromatic oils from the Australian flora. In Proceedings of the International Conference on Medicinal and Aromatic Plants (Part II) 597, Budapest, Hungary, 8–11 July 2001; pp. 79–89.

22. Blewitt, M.; Southwell, I.A. *Backhousia anisata* Vickery, an alternative source of (E)-anethole. *J. Essent. Oil Res.* **2000**, *12*, 445–454. [CrossRef]

23. Brophy, J.J.; Boland, D.J. The leaf essential oil of two chemotypes of *Backhousia anisata* Vickery. *Flavour Fragr. J.* **1991**, *6*, 187–188. [CrossRef]

24. Fujita, K.I.; Fujita, T.; Kubo, I. Anethole, a potential antimicrobial synergist, converts a fungistatic dodecanol to a fungicidal agent. *Phytother. Res.* **2007**, *21*, 47–51. [CrossRef] [PubMed]

25. Himejima, M.; Kubo, I. Fungicidal activity of polygodial in combination with anethole and indole against *Candida albicans*. *J. Agric. Food Chem.* **1993**, *41*, 1776–1779. [CrossRef]

26. Dupont, S.; Caffin, N.; Bhandari, B.; Dykes, G.A. In vitro antibacterial activity of Australian native herb extracts against food-related bacteria. *Food Control.* **2006**, *17*, 929–932. [CrossRef]

27. Sultanbawa, Y. Essential Oils in Food Applications: Australian Aspects. In *Essential Oils in Food Preservation, Flavor and Safety*; Preedy, V.R., Ed.; Elsevier: Amsterdam, The Netherlands, 2016; pp. 155–160.

28. Zhao, J.; Agboola, S.O. *Functional Properties of Australian Bushfoods: A Report for the Rural Industries Research and Development Corporation*; RIRDC: Barton, Australia, 2007.

29. Tiwari, B.K.; Valdramidis, V.P.; O'Donnell, C.P.; Muthukumarappan, K.; Bourke, P.; Cullen, P.J. Application of natural antimicrobials for food preservation. *J. Agric. Food Chem.* **2009**, *57*, 5987. [CrossRef] [PubMed]

30. Savoia, D. Plant-derived antimicrobial compounds: Alternatives to antibiotics. *Future Microbiol.* **2012**, *7*, 979–990. [CrossRef] [PubMed]

31. Aneja, K.R.; Dhiman, R.; Aggarwal, N.K.; Aneja, A. Emerging preservation techniques for controlling spoilage and pathogenic microorganisms in fruit juices. *Int. J. Microbiol.* **2014**, *2014*, 758942. [CrossRef] [PubMed]

32. Ahmad, R.; Ali, A.M.; Israf, D.A.; Ismail, N.H.; Shaari, K.; Lajis, N.H. Antioxidant, radical-scavenging, anti-inflammatory, cytotoxic and antibacterial activities of methanolic extracts of some Hedyotis species. *Life Sci.* **2005**, *76*, 1953–1964. [CrossRef] [PubMed]

33. Shimada, S.; Andou, M.; Naito, N.; Yamada, N.; Osumi, M.; Hayashi, R. Effects of hydrostatic pressure on the ultrastructure and leakage of internal substances in the yeast Saccharomyces cerevisiae. *Appl. Microbiol. Biotechnol.* **1993**, *40*, 123–131. [CrossRef]

34. Singleton, V.; Rossi, J.A. Colorimetry of total phenolics with phosphomolybdic-phosphotungstic acid reagents. *Am. J. Enol. Vitic.* **1965**, *16*, 144–158.

35. Nirmal, N.P.; Panichayupakaranant, P. Antioxidant, antibacterial, and anti-inflammatory activities of standardized brazilin-rich *Caesalpinia sappan* extract. *Pharm. Boil.* **2015**, *53*, 1339–1343. [CrossRef] [PubMed]

36. Karathia, H.; Vilaprinyo, E.; Sorribas, A.; Alves, R. Saccharomyces cerevisiae as a model organism: A comparative study. *PLoS ONE* **2011**, *6*, e16015. [CrossRef] [PubMed]

37. Bakkali, F.; Averbeck, S.; Averbeck, D.; Idaomar, M. Biological effects of essential oils—A review. *Food Chem. Toxicol.* **2008**, *46*, 446–475. [CrossRef] [PubMed]

38. Freiesleben, S.; Jager, A. Correlation between Plant Secondary Metabolites and Their Antifungal Mechanisms—A Review. *Med. Aromat. Plants* **2014**, *3*, 1–6.

39. Hayouni, E.A.; Abedrabba, M.; Bouix, M.; Hamdi, M. The effects of solvents and extraction method on the phenolic contents and biological activities in vitro of Tunisian *Quercus coccifera* L. and *Juniperus phoenicea* L. fruit extracts. *Food Chem.* **2007**, *105*, 1126–1134. [CrossRef]

40. Zafeiropoulou, T.; Evageliou, V.; Gardeli, C.; Yanniotis, S.; Komaitis, M. Retention of trans anethole by gelatine and starch matrices. *Food Chem.* **2010**, *123*, 364–368. [CrossRef]

41. Huang, Y.; Zhao, J.; Zhou, L.; Wang, J.; Gong, Y.; Chen, X.; Guo, Z.; Wang, Q.; Jiang, W. Antifungal activity of the essential oil of *Illicium verum* fruit and its main component trans-anethole. *Molecules* **2010**, *15*, 7558–7569. [CrossRef] [PubMed]

42. Kurekci, C.; Padmanabha, J.; Bishop-Hurley, S.L.; Hassan, E.; Al Jassim, R.A.; McSweeney, C.S. Antimicrobial activity of essential oils and five terpenoid compounds against *Campylobacter jejuni* in pure and mixed culture experiments. *Int. J. Food Microbiol.* **2013**, *166*, 450–457. [CrossRef] [PubMed]

43. Bassolé, I.H.N.; Lamien-Meda, A.; Bayala, B.; Tirogo, S.; Franz, C.; Novak, J.; Nebié, R.C.; Dicko, M.H. Composition and Antimicrobial Activities of *Lippia multiflora* Moldenke, *Mentha x piperita* L. and *Ocimum basilicum* L. Essential Oils and Their Major Monoterpene Alcohols Alone and in Combination. *Molecules* **2010**, *15*, 7825. [CrossRef] [PubMed]

44. Delaquis, P.J.; Stanich, K.; Girard, B.; Mazza, G. Antimicrobial activity of individual and mixed fractions of dill, cilantro, coriander and eucalyptus essential oils. *Int. J. Food Microbiol.* **2002**, *74*, 101–109. [CrossRef]

45. Sultanbawa, Y.; Cusack, A.; Currie, M.; Davis, C. An innovative microplate assay to facilitate the detection of antimicrobial activity in plant extracts. *J. Rapid Methods Autom. Microbiol.* **2009**, *17*, 519–534. [CrossRef]

46. Weerakkody, N.S.; Caffin, N.; Turner, M.S.; Dykes, G.A. In vitro antimicrobial activity of less-utilized spice and herb extracts against selected food-borne bacteria. *Food Control* **2010**, *21*, 1408–1414. [CrossRef]

47. López, V.; Akerreta, S.; Casanova, E.; García-Mina, J.M.; Cavero, R.Y.; Calvo, M.I. In vitro antioxidant and anti-rhizopus activities of *Lamiaceae herbal* extracts. *Plant Foods Hum. Nutr.* **2007**, *62*, 151–155. [CrossRef] [PubMed]

48. Barchan, A.; Bakkali, M.; Arakrak, A.; Pagán, R.; Laglaoui, A. The effects of solvents polarity on the phenolic contents and antioxidant activity of three *Mentha species* extracts. *Int. J. Curr. Microbiol. Appl. Sci.* **2014**, *3*, 399–412.

49. Sakulnarmrat, K.; Konczak, I. Composition of native Australian herbs polyphenolic-rich fractions and in vitro inhibitory activities against key enzymes relevant to metabolic syndrome. *Food Chem.* **2012**, *134*, 1011–1019. [CrossRef] [PubMed]

50. Li, H.; Chen, B.; Yao, S. Application of ultrasonic technique for extracting chlorogenic acid from *Eucommia ulmodies* Oliv. (*E. ulmodies*). *Ultrason. Sonochem.* **2005**, *12*, 295–300. [CrossRef] [PubMed]

51. Du, F.-Y.; Xiao, X.-H.; Luo, X.-J.; Li, G.-K. Application of ionic liquids in the microwave-assisted extraction of polyphenolic compounds from medicinal plants. *Talanta* **2009**, *78*, 1177–1184. [CrossRef] [PubMed]

52. Iloki-Assanga, S.B.; Lewis-Luján, L.M.; Lara-Espinoza, C.L.; Gil-Salido, A.A.; Fernandez-Angulo, D.; Rubio-Pino, J.L.; Haines, D.D. Solvent effects on phytochemical constituent profiles and antioxidant activities, using four different extraction formulations for analysis of *Bucida buceras* L. and *Phoradendron californicum*. *BMC Res. Notes* **2015**, *8*, 396. [CrossRef] [PubMed]

53. Brahmi, F.; Mechri, B.; Dabbou, S.; Dhibi, M.; Hammami, M. The efficacy of phenolics compounds with different polarities as antioxidants from olive leaves depending on seasonal variations. *Ind. Crops Prod.* **2012**, *38*, 146–152. [CrossRef]

54. Vlietinck, A.; Van Hoof, L.; Totte, J.; Lasure, A.; Berghe, D.V.; Rwangabo, P.; Mvukiyumwami, J. Screening of hundred Rwandese medicinal plants for antimicrobial and antiviral properties. *J. Ethnopharmacol.* **1995**, *46*, 31–47. [CrossRef]

55. Shan, B.; Cai, Y.-Z.; Brooks, J.D.; Corke, H. The in vitro antibacterial activity of dietary spice and medicinal herb extracts. *Int. J. Food Microbiol.* **2007**, *117*, 112–119. [CrossRef] [PubMed]

56. Ali-Shtayeh, M.; Yaghmour, R.M.-R.; Faidi, Y.; Salem, K.; Al-Nuri, M. Antimicrobial activity of 20 plants used in folkloric medicine in the Palestinian area. *J. Ethnopharmacol.* **1998**, *60*, 265–271. [CrossRef]

57. Chuang, P.-H.; Lee, C.-W.; Chou, J.-Y.; Murugan, M.; Shieh, B.-J.; Chen, H.-M. Anti-fungal activity of crude extracts and essential oil of *Moringa oleifera* Lam. *Bioresour. Technol.* **2007**, *98*, 232–236. [CrossRef] [PubMed]

58. Ultee, A.; Bennik, M.; Moezelaar, R. The phenolic hydroxyl group of carvacrol is essential for action against the food-borne pathogen *Bacillus cereus*. *Appl. Environ. Microbiol.* **2002**, *68*, 1561–1568. [CrossRef] [PubMed]

59. Santoro, G.F.; das Graças Cardoso, M.; Guimarães, L.G.L.; Salgado, A.P.S.; Menna-Barreto, R.F.; Soares, M.J. Effect of oregano (*Origanum vulgare* L.) and thyme (*Thymus vulgaris* L.) essential oils on *Trypanosoma cruzi* (Protozoa: Kinetoplastida) growth and ultrastructure. *Parasitol. Res.* **2007**, *100*, 783–790. [CrossRef] [PubMed]
60. Prashar, A.; Hili, P.; Veness, R.G.; Evans, C.S. Antimicrobial action of palmarosa oil (*Cymbopogon martinii*) on *Saccharomyces cerevisiae*. *Phytochemistry* **2003**, *63*, 569–575. [CrossRef]
61. Lima, I.O.; Pereira, F.D.O.; Oliveira, W.A.D.; Lima, E.D.O.; Menezes, E.A.; Cunha, F.A.; Diniz, M.D.F.F.M. Antifungal activity and mode of action of carvacrol against *Candida albicans* strains. *J. Essent. Oil Res.* **2013**, *25*, 138–142. [CrossRef]
62. Hoehamer, C.F.; Cummings, E.D.; Hilliard, G.M.; Rogers, P.D. Changes in the proteome of *Candida albicans* in response to azole, polyene, and echinocandin antifungal agents. *Antimicrob. Agents Chemother.* **2010**, *54*, 1655–1664. [CrossRef] [PubMed]
63. Miron, D.; Battisti, F.; Silva, F.K.; Lana, A.D.; Pippi, B.; Casanova, B.; Gnoatto, S.; Fuentefria, A.; Mayorga, P.; Schapoval, E.E. Antifungal activity and mechanism of action of monoterpenes against dermatophytes and yeasts. *Rev. Bras. Farmacogn.* **2014**, *24*, 660–667. [CrossRef]

 *foods*

Article

Influence of *Clitoria ternatea* Flower Extract on the In Vitro Enzymatic Digestibility of Starch and Its Application in Bread

Charoonsri Chusak [1], Christiani Jeyakumar Henry [2,3], Praew Chantarasinlapin [1], Varanya Techasukthavorn [1] and Sirichai Adisakwattana [1,*]

[1] Department of Nutrition and Dietetics, Faculty of Allied Health Sciences, Chulalongkorn University, Bangkok 10330, Thailand; charoonsri.c@gmail.com (C.C.); praewchan@yahoo.com (P.C.); varanya.te@gmail.com (V.T.)
[2] Clinical Nutrition Research Centre, Singapore Institute for Clinical Sciences (SICS), Agency for Science, Technology and Research, Singapore 117599, Singapore; jeya_henry@sics.a-star.edu.sg
[3] Department of Biochemistry, Yong Loo Lin School of Medicine, National University of Singapore, Singapore 117596, Singapore
* Correspondence: Sirichai.a@chula.ac.th; Tel: +66-2-218-1099 (ext. 111)

Received: 21 May 2018; Accepted: 29 June 2018; Published: 2 July 2018

Abstract: This study aimed to assess the effect of the *Clitoria ternatea* L. flower extract (CTE), on the inhibition of pancreatic α-amylase, in vitro starch hydrolysis, and predicted the glycemic index of different type of flours including potato, cassava, rice, corn, wheat, and glutinous rice flour. The application in a bakery product prepared from flour and CTE was also determined. The results demonstrated that the 1% and 2% (w/v) CTE inhibited the pancreatic α-amylase activity by using all flours as a substrate. Moreover, 0.5%, 1%, and 2% (w/v) CTE showed a significant reduction in the glucose release, hydrolysis index (HI), and predicted glycemic index (pGI) of flour. In glutinous rice flour, 1% and 2% (w/v) CTE had a significantly lower level of rapidly digestible starch (RDS) and slowly digestible starch (SDS) with a concomitant higher level of undigested starch. The statistical analysis demonstrated strong positive significant correlations between the percentage of CTE and the undigested starch of wheat and cassava. The addition of 5%, 10%, and 20% (w/w) CTE significantly reduced the rate of starch digestion of the wheat bread. The pGI of bread incorporated with 5% CTE (w/w) was significantly lower than that of the control bread. Our findings suggest that CTE could reduce the starch digestibility, the HI, and pGI of flour through the inhibition of carbohydrate digestive enzymes. Taken together, CTE may be a potent ingredient for the reduced glycemic index of flours.

Keywords: *Clitoria ternatea* L. flower extract; in vitro starch digestibility; hydrolysis index; predicted glycemic index

---

1. Introduction

Carbohydrates are one of three basic macronutrients that produce energy for our body. Nowadays, edible flour enriched with carbohydrates has been made from several parts of plants, including roots, seeds, and germs [1]. Flour is a common ingredient used for foods and desserts, according to different cooking purposes. The excessive and chronic consumption of flour markedly increases the postprandial blood glucose level and leads to excess visceral fat, which increases both insulin resistance and inflammation, and predisposes one to diabetes, hypertension, and cardiovascular diseases [2]. It has been shown that various types of flour contribute to a different rate and degree of starch hydrolysis, resulting in varying degrees of postprandial blood glucose rise [3]. Normally, the glycemic index (GI) is the measure of the immediate effect on the postprandial glucose level after food consumption, by comparing the percentage of incremental glucose area under the curve (iAUC) of a test food with reference to a standard food. However, the in vivo measurement of GI requires the recruitment of human subjects under ethical committee approval, financial supports, and it is time consuming, and all of these reasons led to widespread acceptance of in vitro starch digestibility studies [4].

The predicted glycemic index (pGI) is a common technique used to measure the rate of carbohydrate hydrolysis in foods [4]. It has been found that in vitro methods used to classify foods based on their digestion characteristics are similar to the in vivo situation [5]. There is a positive correlation between the in vitro and in vivo glycemic response [4]. In the focus of nutritional aspects, carbohydrate foods with a low pGI value (<55) can be considered as beneficial foodstuff for human health, in terms of the prevention and treatment of the metabolic syndromes, diabetes, and cardiovascular diseases [6,7]. Replacing or mixing flours with other ingredients such as fruits and vegetables is one of the alternative approaches to reduce the pGI in carbohydrate foods [8]. For example, pomelo containing polyphenols incorporated into bread could lower the predicted glycemic index probably by inhibiting the activity of the carbohydrate hydrolyzing enzymes [9]. It is becoming clear that plant-based ingredients containing polyphenols delay the action of carbohydrate digestive enzymes and thereby reduce the absorption rate of glucose [10,11]. Previously, polyphenols from the extracts of pomegranate, cranberry, grape, and cocoa could bind to the digestive enzymes (α-amylase and glucoamylase), resulting in the inhibition of starch hydrolysis [12]. Therefore, the addition of plant-based ingredients may be capable of reducing the glycemic index during starch hydrolysis.

Clitoria ternatea L., commonly known as Butterfly pea, is a plant species belonging to the Fabaceae family. This plant is widely distributed in tropical zones such as Asia, the Caribbean, and Central and South America. In traditional Ayurvedic medicine, *Clitoria ternatea* L. has been used for treating stress and depression and enhancing memory [13]. There have been many pharmacological activities reported for this plant, such as antidiabetic [14], antipyretic [15], anti-inflammatory [16], and antimicrobial activity [17]. *Clitoria ternatea* has been reported to contain rutin, kaempferol, delphinidin, and related glycosides [18]. Our recent reports demonstrated that an aqueous extract of CTE inhibited the activity of carbohydrate digestive enzymes such as intestinal α-glucosidase and pancreatic α-amylase [19]. In addition to the biological pigment, the flower of *Clitoria ternatea* has been used as a colorant in various foods, beverages, and desserts in Asia. This colorant flower is regularly mixed with rice, bread, cookies, flours, and other traditional foods and desserts with a variety of ratios. For example, some traditional Thai desserts are made using cassava flour or glutinous rice flour mixed with butterfly pea juice at various concentrations to color and are then steamed until cooked. Moreover, sticky rice noodles are made by the mixture of rice flour with CTE juice. Although the pancreatic α-amylase and α-glucosidase inhibitory activity of CTE is well-documented, studies regarding its effect of pancreatic α-amylase action and in vitro starch digestibility using various types of flour have not been taken. Particularly, the potential food application of CTE in flour-based products remains unknown. Therefore, the aim of the present study was to investigate the effect of *Clitoria ternatea* L. flower extract on the activity of pancreatic α-amylase, in vitro starch hydrolysis, and predicted glycemic index of potato, cassava, rice, corn, wheat, and glutinous rice flour. The application in bread prepared from wheat flour and CTE was also determined.

2. Materials and Methods

2.1. Chemicals and Reagents

Commercial flours including potato, rice, glutinous rice, wheat, corn, and cassava flours were purchased from a supermarket. Porcine pancreatic α-amylase Type VI-B (catalogue number: A3176) and 3,5-dinitrosalicylic acid were purchased from the Sigma-Aldrich Chemical Co., Ltd. (St. Louis, MO, USA). Amyloglucosidase was obtained from Roche Diagnostics (Indianapolis, IN, USA). The glucose oxidase-peroxidase (GOPOD) kit was purchased from HUMAN GmbH (Wiesbaden, Germany).

2.2. Extraction

The dried flower of *Clitoria ternatea* L. was purchased from a local herbal drug store, Bangkok, Thailand. The extraction was performed according to a previous study [20]. The content of the phenolic compounds and the total anthocyanins in the CTE was 53.08 ± 0.08 mg gallic acid equivalents/g extract and 1.08 ± 0.12 mg delphinidin-3-glucoside equivalents/g extract, respectively.

2.3. Preparation of Flour and Extract

Flour (0.25 g) was dissolved with 50 mL of boiled water at 100 °C and stirred for 10 min. The flour solution was allowed to cool at room temperature for 10 min. Furthermore, the CTE powder was dissolved in 0.1 M phosphate buffer saline (PBS) or 0.2 M sodium acetate buffer. The CTE solution was vortexed for 10 min. The CTE solution was added into the flour solution (final concentration: 0.5%, 1%, and 2% w/v) and subjected to in vitro digestion.

2.4. Inhibition of Pancreatic α-Amylase

The activity of pancreatic α-amylase was carried out using a modified procedure of Adisakwattana et al. [19]. Fifty microliters of the flour solution were mixed with 100 μL of CTE in 0.1 M phosphate buffer saline (PBS). Fifty microliters of porcine pancreatic α-amylase (15 U/mL) in 0.1 M PBS, pH 6.9, was then added and the mixture was made to 250 μL with 0.1 M PBS. After incubation at 37 °C for 10 min, the reaction was terminated by adding 250 μL of DNS reagent (1% DNS, 0.2% phenol, 0.05% Na_2SO_3, and 1% NaOH in distilled water) and heated at 100 °C for 10 min. Then, 40% potassium sodium tartrate (250 μL) was added to stabilize the color. After cooling at room temperature, the absorbance was measured at 540 nm. The pancreatic α-amylase inhibitory activity was calculated as the percentage inhibition. A control was prepared using the same procedure, replacing the CTE solution with 0.1 M PBS.

2.5. In Vitro Starch Digestibility and Predicted Glycemic Index (pGI)

The in vitro digestion of flour and CTE was performed according to a previous method with some modifications [21,22]. Fifty microliters of the flour solution were mixed with 100 μL of CTE in 0.2 M sodium acetate buffer. The mixture was incubated with 50 μL of porcine pancreatic α-amylase (15 U/mL) and 50 μL of amyloglucosidase (31.25 μg/mL) in 0.2 M sodium acetate buffer, pH 6.0, at 37 °C for 180 min. After heating at 100 °C for 10 min, for stopping the reaction, the supernatant was measured for the glucose content using a glucose oxidase-peroxidase (GOPOD) kit. The values were plotted a graph and the area under the curve (AUC) was calculated using the trapezoidal rule. The hydrolysis index (HI) was calculated from the percentage of the area under the hydrolysis curve of the sample to the area under the curve of the standard glucose. The predicted glycemic indices (pGI) of the samples were estimated according to the followed equation: pGI = 39.71 + 0.549 HI [4]. A control was prepared using the same procedure, replacing the CTE solution with 0.2 M sodium acetate buffer.

2.6. Estimation of Starch Fraction

The starch fraction was calculated based on the in vitro starch digestibility of samples. Rapidly digestible starch (RDS) was calculated as the amount of glucose present in the sample at 20 min of the in vitro digestion, whereas slowly digestible starch (SDS) was calculated as the difference between the amount of glucose measured at 120 min and 20 min [21,23]. The undigested starch was calculated as the amount of glucose that was not digested within 120 min. The conversion factor from the glucose to starch was 0.9.

2.7. Bread Preparation

Wheat flour and other dry ingredients as a % on the weight of flour basis, including sugar (5%), salt (1.8%), yeast (3%), Benecel methylcellulose, and hydroxypropylmethylcellulose (2%) (Ashland,

Covington, KY, USA), were mixed using a Kitchen-Aid bowl mixer at speed 53 rpm for 1 min. The levels of CTE incorporated into this formulation were 5%, 10%, and 20% (w/w) of the wheat flour basis, by adding to mix with other dry ingredients. Based on the wheat flour, vegetable oil (6%), white egg (10%), and milk (70%) were added and then mixed together at speed of 160 rpm for 10 min. After that, the batter was poured into a mold, placed at 30 °C and 90% relative humidity for 50 min, and baked at 150 °C for 40 min. The loaf was removed from the mold and cooled at room temperature. The bread sample was packed in a sealed polyethylene bag until analysis. After baking, the in vitro starch hydrolysis of the bread with or without CTE was determined to indicate the in vitro starch digestibility and pGI. The bread samples were weighed (1 g) and mixed with distilled water (10 mL) while stirring for 10 min. The in vitro starch digestion of the bread was performed according to the above-mentioned method. A control bread was prepared using the same procedure, without the CTE.

2.8. Statistical Analysis

The data were expressed as mean ± standard error of the mean (SEM). The statistical significance of the results was evaluated by one-way ANOVA using Duncan multiple comparisons, using SPSS version 22.0 (SPSS Inc., Chicago, IL, USA). The Pearson's correlation coefficients (r) were calculated between the CTE concentration and undigested starch. $p < 0.05$ was considered statistically significant.

3. Results

3.1. Inhibition of Pancreatic α-Amylase

As shown in Figure 1, the amount of maltose released from all of the flours was observed after 10 min of incubation. There was a significant reduction for the release of maltose after mixing the potato, rice, glutinous rice, wheat, corn, and cassava flour with 1% and 2% (w/v) CTE, compared with the control ($p < 0.05$).

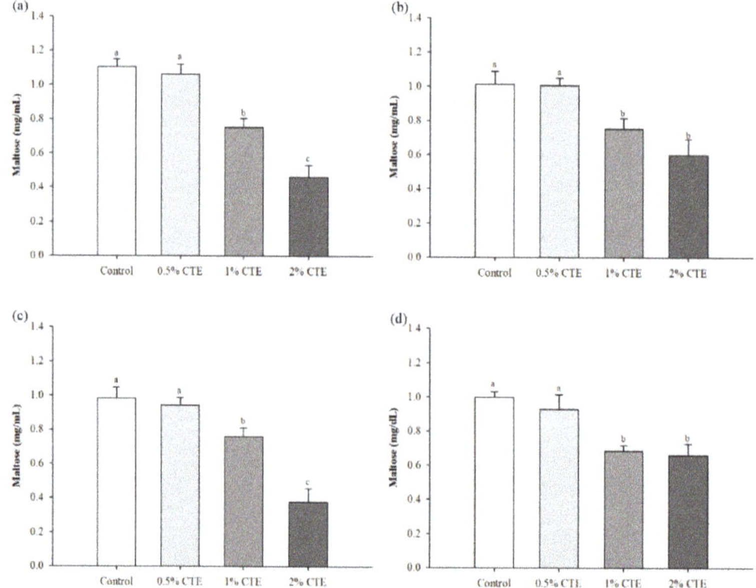

Figure 1. *Cont.*

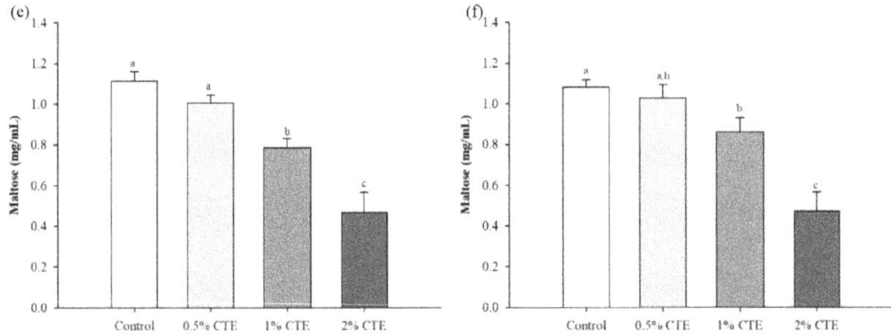

Figure 1. The amount of maltose released from (**a**) potato flour, (**b**) rice flour, (**c**) glutinous rice flour, (**d**) wheat flour, (**e**) corn flour, and (**f**) cassava flour when combination with the different concentrations of CTE against pancreatic α-amylase activity at 10 min. The results are expressed as mean $\pm$ standard error of the mean (SEM), $n = 4$. The different letters denote statistically significant differences in mean values. ($p < 0.05$) Mean values with the same superscript letters (a or b) were similar and no statistically significant differences were observed for these samples. CTE—*Clitoria ternatea* L. flower extract.

The percentage of pancreatic α-amylase inhibitory activity after mixing the CTE into flour is shown in Table 1. The increased percentage of pancreatic α-amylase inhibitory activity was concomitant with the increased concentration of CTE. The results demonstrated that the mixture of potato, rice, glutinous rice, wheat, corn, and cassava flours with 1% and 2% (w/v) CTE resulted in a higher pancreatic α-amylase inhibitory activity than that of the 0.5% (w/v) of CTE ($p < 0.05$). At 2% (w/v) CTE, the potato flour had the highest percentage of pancreatic α-amylase inhibitory activity, followed by glutinous rice, rice, wheat, corn, and cassava, respectively.

Table 1. The percentage of pancreatic α-amylase inhibitory activity of CTE.

CTE	% Inhibition					
	Potato	Rice	Glutinous Rice	Wheat	Corn	Cassava
0.5% (w/v)	7.1 $\pm$ 4.7	23.8 $\pm$ 9.2	20.8 $\pm$ 9.2	17.4 $\pm$ 6.5	24.1 $\pm$ 9.5	12.9 $\pm$ 1.4
1% (w/v)	56.7 $\pm$ 7.8	82.1 $\pm$ 8.1	51.7 $\pm$ 8.5	50.1 $\pm$ 7.7	48.3 $\pm$ 8.2	34.5 $\pm$ 8.9
2% (w/v)	93.4 $\pm$ 5.7	87.3 $\pm$ 13.4	89.7 $\pm$ 11.9	85.2 $\pm$ 4.3	81.6 $\pm$ 10.6	79.9 $\pm$ 19.7

The results are expressed as mean $\pm$ standard error of the mean (SEM), $n = 4$. CTE—*Clitoria ternatea* L. flower extract.

3.2. In Vitro Starch Digestibility and Predicted Glycemic Index (pGI)

The results of in vitro starch digestibility of flour at different concentration of CTE are shown in Figure 2. The incorporation of CTE decreased the glucose released from flour. The results demonstrated that the amount of glucose released from the mixture of CTE and flours was lower than the control. The addition of 2% (w/v) CTE into flour significantly caused the highest inhibition of starch digestibility.

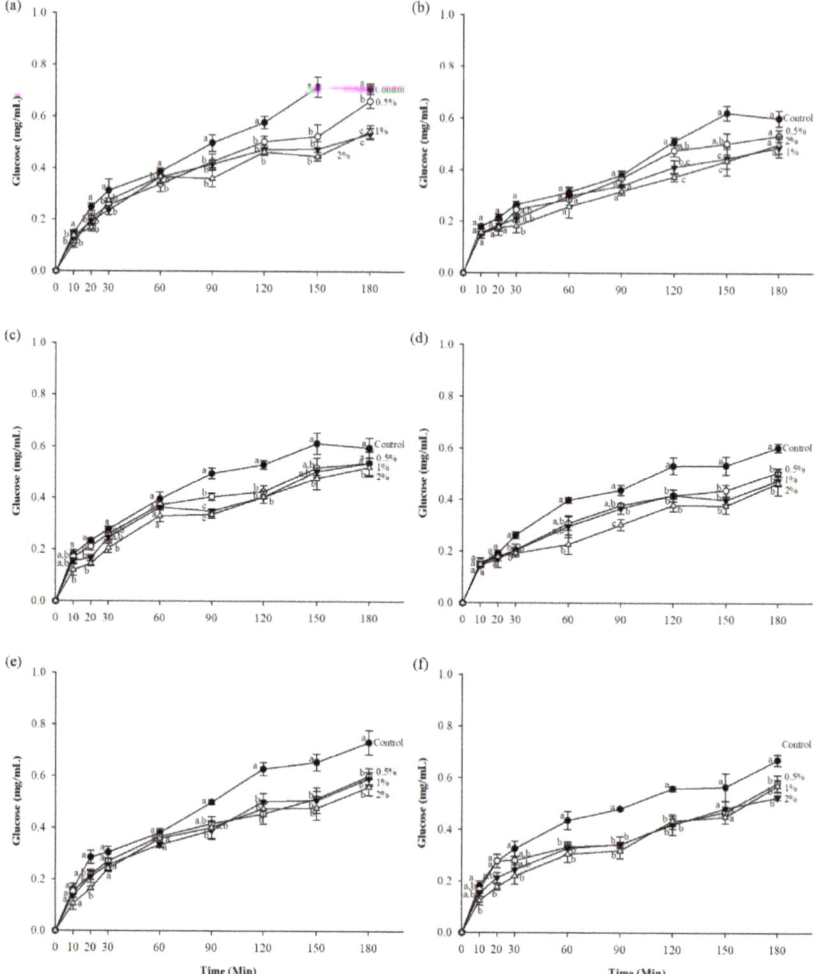

Figure 2. The amount of glucose released from (**a**) potato flour, (**b**) rice flour, (**c**) glutinous rice flour, (**d**) wheat flour, (**e**) corn flour, and (**f**) cassava flour when in combination with the different concentrations of CTE under in vitro digestibility during 180 min. The value of 0.5%, 1%, and 2% (*w/v*) represent the concentration of CTE, respectively. The results are expressed as mean ± standard error of the mean (SEM), *n* = 4. The different letters denote statistically significant differences in mean values. (*p* < 0.05) Mean values with the same superscript letters (a, b or c) were similar and no statistically significant differences were observed for these samples.

Table 2 represents an interpretation of the hydrolysis index (HI) and the predicted glycemic index (pGI) of all of the samples. The HI and pGI of potato, rice, glutinous rice, wheat, corn, and cassava flour with 0.5%, 1%, and 2% (*w/v*) of CTE were significantly lowered when compared with the control (*p* < 0.05). Interestingly, the addition of 2% (*w/v*) CTE caused the reduction of pGI of the glutinous rice, wheat, and cassava flour from the high value to the medium value (GI < 70).

Table 2. The hydrolysis index (HI) and predicted glycemic index (pGI) of the flours with CTE.

CTE	Hydrolysis Index (HI)					
	Potato	**Rice**	**Glutinous Rice**	**Wheat**	**Corn**	**Cassava**
Control	85.2 ± 0.5 [a]	74.1 ± 4.3 [a]	86.9 ± 2.4 [a]	74.3 ± 2.5 [a]	94.4 ± 3.6 [a]	81.0 ± 1.3 [a]
0.5% (*w/v*)	71.0 ± 1.6 [b]	63.4 ± 2.4 [a,b]	65.5 ± 4.1 [b]	62.4 ± 3.7 [b]	69.5 ± 3.2 [b]	65.5 ± 2.2 [b]
1% (*w/v*)	68.8 ± 3.7 [b,c]	57.2 ± 4.9 [b]	63.7 ± 5.2 [b]	61.0 ± 1.4 [b]	68.5 ± 2.2 [b]	61.3 ± 3.4 [b,c]
2% (*w/v*)	62.2 ± 2.1 [c]	55.2 ± 3.3 [b]	50.3 ± 5.1 [c]	50.4 ± 4.8 [c]	59.0 ± 2.3 [c]	51.5 ± 6.0 [c]

CTE	Predicted Glycemic Index (pGI)					
	Potato	**Rice**	**Glutinous Rice**	**Wheat**	**Corn**	**Cassava**
Control	86.5 ± 0.3 [a]	80.4 ± 2.4 [a]	87.2 ± 1.3 [a]	80.5 ± 1.4 [a]	91.2 ± 2.0 [a]	84.2 ± 0.7 [a]
0.5% (*w/v*)	78.7 ± 0.9 [b]	74.5 ± 1.3 [a,b]	75.7 ± 2.2 [b]	74.0 ± 2.0 [b]	77.9 ± 1.8 [b]	75.7 ± 1.2 [b]
1% (*w/v*)	77.5 ± 2.0 [b,c]	71.1 ± 2.7 [b]	74.7 ± 2.8 [b]	73.2 ± 0.8 [b]	77.3 ± 1.2 [b]	73.4 ± 1.9 [b,c]
2% (*w/v*)	73.8 ± 1.1 [c]	70.0 ± 1.8 [b]	67.4 ± 2.8 [c]	67.4 ± 2.6 [c]	72.1 ± 1.3 [c]	68.0 ± 3.3 [c]

The results are expressed as mean $\pm$ standard error of the mean (SEM), $n = 4$. The different letters denote statistically significant differences in mean values. ($p < 0.05$) Mean values with the same superscript letters (a, b, or c) were similar and no statistically significant differences were observed for these samples.

3.3. Starch Fraction

The RDS, SDS, and undigested starch of flours are presented in Figure 3. The RDS content of six flours mixed with at 2% (*w/v*) CTE significantly decreased when compared with the control ($p < 0.05$). The addition of CTE at 0.5%, 1%, and 2% (*w/v*) caused a reduction in the SDS content of glutinous rice, corn, and cassava flour ($p < 0.05$). However, the CTE did not alter the SDS content of potato flour. The observed results also found that only the glutinous rice flour significantly increased the undigested starch with the addition of CTE ($p < 0.05$). Table 3 shows the correlation between the concentration of CTE and undigested starch of flour. The undigested starch of the wheat and cassava flour correlated significantly and positively with the concentration of CTE ($r = 0.650$ and 0.758, respectively; $p < 0.05$). However, no significant correlation was observed between the concentration of CTE and other flours, including potato flour, rice flour, glutinous rice flour, and corn flour.

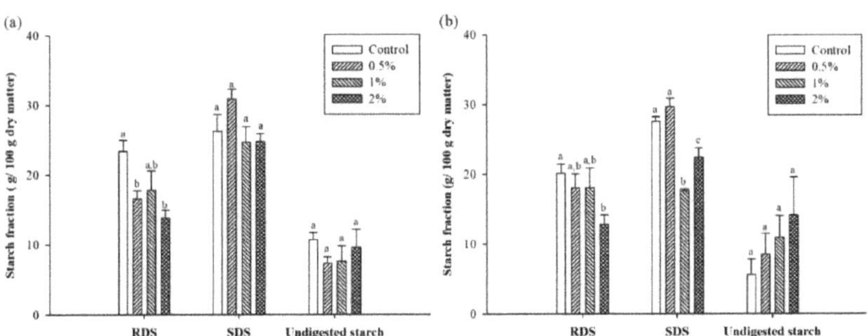

Figure 3. *Cont.*

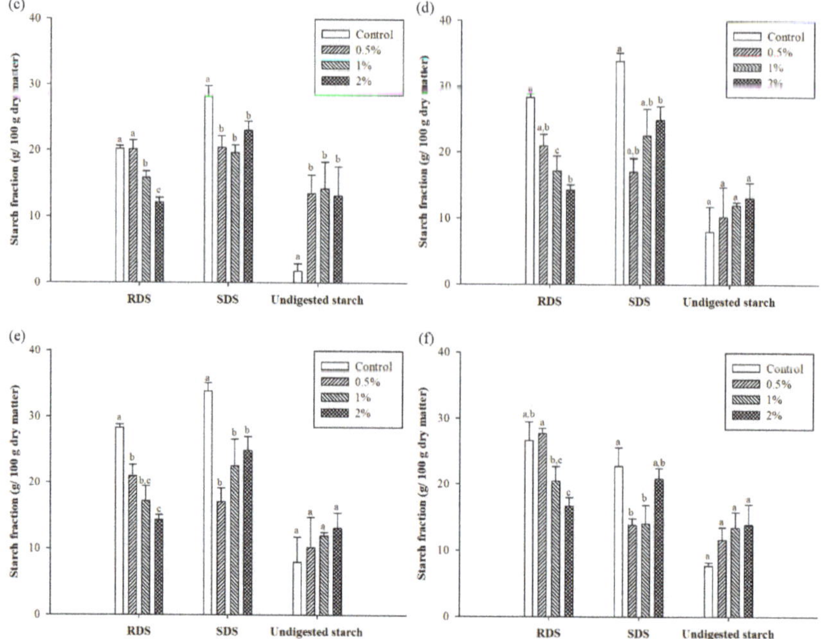

Figure 3. Starch fraction after in vitro digestibility of (**a**) potato flour, (**b**) rice flour, (**c**) glutinous rice flour, (**d**) wheat flour, (**e**) corn flour, and (**f**) cassava flour when in combination with the different concentration of CTE. The value of 0.5%, 1%, and 2% (*w/v*) represent the concentrations of CTE, respectively. The results are expressed as mean ± standard error of the mean (SEM), *n* = 4. The different letters denote statistically significant differences in mean values. (*p* < 0.05) Mean values with the same superscript letters (a, b or c) were similar and no statistically significant differences were observed for these samples. RDS: rapidly digestible starch; SDS: slowly digestible starch.

Table 3. Correlation coefficients calculated between the concentrations of *Clitoria ternatea* (CTE) and undigested starch contents after in vitro digestion.

	Potato	Rice	Glutinous Rice	Wheat	Corn	Cassava
CTE	−0.040	0.511	0.486	0.650 *	0.373	0.758 *

* Significant correlations (*p* < 0.05).

3.4. In Vitro Digistibility of Bread

Cross sections of bread made from wheat flour and CTE are shown in Figure 4. The in vitro starch digestibility of wheat bread with 5%, 10%, and 20% (*w/w*) of the CTE are presented in Figure 5a. The amount of glucose released from the bread with CTE was lower than that of the control. The addition of 5%, 10%, and 20% (*w/w*) CTE into wheat bread significantly reduced the rate of starch digestion after 120, 150, and 180 min of incubation (*p* <0.05). As shown in Figure 5b, the iAUCs for the glucose release of bread incorporated with 5–20% (*w/w*) CTE were 8136 ± 82, 8997 ± 42, and 7363 ± 386 mg/dL·min, respectively (the control = 11,364 ± 172 mg/dL·min). The pGI of the bread with 5–20% (*w/w*) CTE was 65.40 ± 0.26, 68.11 ± 0.13, and 62.96 ± 1.22, respectively, whereas the wheat bread had the pGI of 75.58 ± 0.54.

Figure 4. The cross section of bread made from wheat flour and CTE.

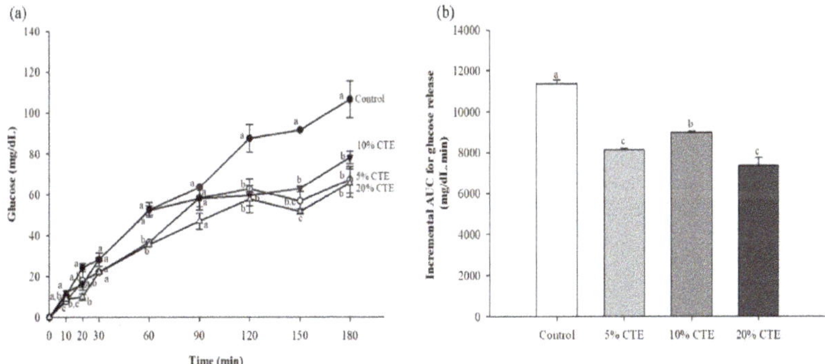

Figure 5. The amount of glucose released from wheat bread after in vitro digestibility (**a**) and incremental area under the curve (iAUC) for glucose release (**b**) when in combination with the different concentrations of CTE. The values of 5%, 10%, and 20% (*w/w*) represent the concentration of CTE, respectively. The results are expressed as mean ± standard error of the mean (SEM), *n* = 4. The different letters denote statistically significant differences in mean values. (*p* < 0.05) Mean values with the same superscript letters (a, b or c) were similar and no statistically significant differences were observed for these samples.

4. Discussion

Starchy foods or ingredients are digested by amylolytic α-amylases and α-glucosidase enzymes, including maltase-glucoamylase and sucrase-isomaltase, at the brush border of the small intestine [24]. Thereafter, the absorbable glucose is transported into the bloodstream through the glucose transporter in the small intestine. The high rate of digestion and absorption of these foods contributes to a rise in postprandial glucose, related to health consequences. The physio-chemical properties of carbohydrate foods are normally investigated by measurement of the rate and extent of glucose release after enzymatic digestion under controlled conditions [25]. Our findings demonstrated that a higher amount of maltose and glucose released from flours was observed after in vitro digestion. When the CTE was mixed with the flours, the release of maltose and glucose was significantly decreased. These findings suggest that CTE has a potential to reduce the release of maltose and glucose from flours, leading to a delay in the rate of starch digestibility. In agreement with another study, CTE inhibited the pancreatic α-amylase and intestinal α-glucosidase related to its phytochemical compounds [26]. It has been revealed that the phytochemical compounds in CTE are delphinidin-3, 5-glucoside, delphinidin-3-glucoside, malvidin-3-glucoside, delphinidin-based ternatins (ternatins A1–A3, B1–B4, C1–C5, and D1–D3), kaempferol, quercetin-3-*O*-(2-rhamnosyl) rutinoside, and rutin [18]. In particular, rutin and kaempferol could inhibit the pancreatic α-amylase and intestinal α-glucosidase activity [27,28]. Moreover, the natural delphinidin and malvidin compounds have shown a competitive inhibiting effect against the intestinal α-glucosidase [29,30]. A study conducted by Podsędek et al. [31]

observed that the degree of the inhibitory effect on the carbohydrate digestive enzymes is positively correlated with the concentration of anthocyanins. We suggest that the phytochemical compounds in CTE may contribute to delaying the hydrolysis of starch by inhibiting the carbohydrate digestive enzymes, including pancreatic α-amylase and intestinal α-glucosidase. Additionally, Zhu et al. [32] also explained other mechanisms of polyphenols on delayed starch digestibility. It found that polyphenol could interact with the starch chains to form the complex, resulting in the alteration of enzyme susceptibility. This evidence was supported from the higher content of undigested starch after in vitro digestion of starch and anthocyanins in blue maize [33]. The interaction between polyphenols and starch was due to the non-covalent bonding and/or the hydrogen binding formation [34]. Further studies are needed to define the hypothesized mechanisms specific to the interaction between the anthocyanins in the CTE and flours, and the type of enzyme inhibition.

For nutritional purposes, the starch in food is generally classified into three categories, including rapidly digestible starch (RDS), slowly digestible starch (SDS), and resistant starch (RS) [35]. In terms of RDS, starch is readily and completely digested in the small intestine associated with more rise postprandial plasma glucose within first 20 min. SDS is a complete digestion of starch in the small intestine with slow rate [5]. Furthermore, RS is defined as the dietary starch that resists digestion in the small intestine. According to a previous study [36], undigested starch contributes to the resistant starch, which is due to the inhibitory activity of α-amylase by antinutrients (e.g., polyphenols). The various amount of starch fractions in the different types of flour depends on their physical and chemical characteristics [37]. The observed results indicate that all flours had a higher level of RDS with a concomitant lower content of undigested starch. The gelatinization is one of the factors affecting starch hydrolysis during the cooking process [38]. Thus, the gelatinization of the starch granule induces an increase in the RDS response with the release of the glucose molecule. The results also showed that the flour with a high RDS content produced a significantly higher level of HI and pGI, whereas the higher SDS content with lower RDS reduced the levels of HI and pGI. A previous study supported our findings, indicating that the pGI value was correlated with the parameters of the digestible starch fractions, including RDS and SDS. In particular, SDS is found to be the main contributing factor to the GI [39]. It has been reported that an intake of diet containing a high RDS level could induce a rapid hyperglycemic response and a subsequent glucose-induced insulin secretion from pancreatic β-cells [5]. In contrast, undigested starch (RS) in human diets provides functional properties and applications for delaying postprandial glucose [40] and improving postprandial insulin [41]. Our findings demonstrated that the addition of CTE into glutinous rice, wheat, corn, and cassava flour causes the reduction of RDS and SDS, in relation to the increased content of undigested starch. Moreover, the undigested starch of wheat and cassava flours significantly and strongly correlated with the concentration of CTE. For traditional use, glutinous rice flour, ground from glutinous rice or sticky rice, is usually used as the ingredient for desserts, sweets, rice cakes, and puffed rice in Asia and Southeast Asia [42]. Basically, glutinous rice has been classified as high GI because of its high amylopectin content and rate of digestion [43]. The starch digestibility of glutinous rice produces more rapidly and is more complete than other high-amylose rice varieties. Chan et al. [44] found an increased glycemic response and GI values in Caucasian and Asian populations after the consumption of glutinous rice, which was similar to a previous study of Ranawana et al. [45]. In the current study, the mixture of CTE into glutinous rice flour can reduce the pGI value, suggesting that CTE suppresses the digestive process of glutinous rice flour to absorbable monosaccharides. A combination of glutinous rice flour and CTE might have opportunities for flour applications to reduce the GI of the food products.

Several studies have reported that the plant-based diets containing polyphenols alter the glycemic index of various foods. The current study found that the addition of CTE caused the rate of carbohydrate digestion and pGI of wheat bread to slow down. Our results are in agreement with Reshmi et al. [9], who reported the in vitro glycemic impact of bread fortified with pomelo fruit. Because of the action of phenolic compounds and the flavonoids in pomelo, the bread fortification with pomelo caused a lower level of digestible starch with a concomitant increased level of undigested starch. Lemlioglu-Austin et al. [46] also found that the incorporation of phenolic-rich sorghum bran extract into porridges contributes to slow starch digestion with a reduced GI and increased undigested starch. In addition, thermal processing also affects starch digestibility by the alteration of its granular structure [47]. The cooking process with heating and excess water induces the gelatinization of the starch granule, increasing in starch digestibility. The crumb portions of baked bread increased the starch digestibility when compared with the crust portion, because the starches in the crust portion are not completely gelatinization after baking. The fortification of green tea polyphenol in baked bread reduced in glucose release for both the crumb and crust after in vitro digestion [48] and reduced the rapidly digestible starch of white bread samples [49]. During baking, the interaction between gelatinized starch granules and the gluten network occurs in crumb, causing a loss of kinetic energy and a subsequent increase in firmness [50]. Previous evidence revealed that polyphenols can form a complex with bread ingredients including protein and polysaccharides [50]. The formation of polyphenols and polysaccharides or protein as enzymes clearly indicates a reduced in vitro digestibility [34]. Moreover, polyphenols affect the breadmaking quality by altering the flour protein properties [51]. The interaction of polyphenols and gluten proteins in wheat bread is associated with a reduction in protein cross-linking, resulting in decreased bread volume. Our study showed that the addition of CTE caused the bread volume to and air pocket in wheat bread to reduce. These findings are consistent with a study of Pathak et al., who reported that the addition of mango fruit peel powder could decrease the volume and height of loaf, whereas it increased the density of the loaf with less visible air pockets of bread, owning to the compact crumb structure [52]. It could be explained that the gluten network was not completely formed, leading to the ineffectiveness to hold air during fermentation, which caused the decreased loaf volume [53]. According to our findings, the addition of CTE into various types of flours successfully altered the parameters of starch digestibility and consequently decreased the level of HI and pGI. Further studies are warranted to elucidate whether the consumption of bread incorporated with CTE delays postprandial glucose in humans.

5. Conclusions

The present study demonstrated that CTE could inhibit pancreatic α-amylase activity, leading to a reduction of maltose release from flour. The addition of CTE to flour alters in vitro starch digestibility, resulting in a reduction in the amount of glucose released from the various types of flour and a subsequent reduction of HI and pGI. Moreover, the addition of the CTE reduced the starch digestibility rate and pGI of wheat bread. The findings indicate that the addition of CTE into flour can inhibit the starch digestibility of flour through the inhibition of carbohydrate digestive enzymes, including pancreatic α-amylase and intestinal α-glucosidase. We suggest that CTE may be a potent ingredient for the reduced glycemic index of flours.

Author Contributions: S.A. and C.J.H. were responsible for the study concept and design, supervising the study, and drafting and revising the content of the manuscript. C.C. conducted the study, data collection and analysis, and interpreted the data and drafted the manuscript. P.C. and V.T. were responsible for drafting the content of the manuscript. All of the authors read and approved the final manuscript.

Acknowledgments: We would like to thank the 100th Anniversary Chulalongkorn University Fund for the Doctoral Scholarship, Overseas Research Experience Scholarship for Graduate Student, and the 90th Anniversary Chulalongkorn University Fund (Ratchadaphiseksomphot Endowment Fund), Graduate school, Chulalongkorn University. This research was supported by the Grant for International Research Integration: Chula Research Scholar, and the Grant for Join Funding, Ratchadaphiseksomphot Endowment Fund, Chulalongkorn University. The author gratefully acknowledges Lalana Thiranusornkij and Parichart Thamnarathip from KCG Excellence Center, KCG Corporation Co., Ltd. for providing bread preparation.

Conflicts of Interest: The authors have declared no conflict of interest.

References

1. Lafiandra, D.; Riccardi, G.; Shewry, P.R. Improving cereal grain carbohydrates for diet and health. *J. Cereal Sci.* **2014**, *59*, 312–326. [CrossRef] [PubMed]
2. O'Keefe, J.H.; Gheewala, N.M.; O'Keefe, J.O. Dietary strategies for improving post-prandial glucose, lipids, inflammation, and cardiovascular health. *J. Am. Coll. Cardiol.* **2008**, *51*, 249–255. [CrossRef] [PubMed]
3. Eleazu, C.O. The concept of low glycemic index and glycemic load foods as panacea for type 2 diabetes mellitus; prospects, challenges and solutions. *Afr. Health Sci.* **2016**, *16*, 468–479. [CrossRef] [PubMed]
4. Goñi, I.; Garcia-Alonso, A.; Saura-Calixto, F. A starch hydrolysis procedure to estimate glycemic index. *Nutr. Res.* **1997**, *17*, 427–437. [CrossRef]
5. Jenkins, D.; Wolever, T.; Taylor, R.H.; Barker, H.; Fielden, H.; Baldwin, J.M.; Bowling, A.C.; Newman, H.C.; Jenkins, A.L.; Goff, D.V. Glycemic index of foods: A physiological basis for carbohydrate exchange. *Am. J. Clin. Nutr.* **1981**, *34*, 362–366. [CrossRef] [PubMed]
6. O'dea, K.; Snow, P.; Nestel, P. Rate of starch hydrolysis in vitro as a predictor of metabolic responses to complex carbohydrate in vivo. *Am. J. Clin. Nutr.* **1981**, *34*, 1991–1993. [CrossRef] [PubMed]
7. Englyst, H.N.; Veenstra, J.; Hudson, G.J. Measurement of rapidly available glucose (RAG) in plant foods: A potential in vitro predictor of the glycaemic response. *Br. J. Nutr.* **1996**, *75*, 327–337. [CrossRef] [PubMed]
8. Dewettinck, K.; Van Bockstaele, F.; Kühne, B.; Van de Walle, D.; Courtens, T.; Gellynck, X. Nutritional value of bread: Influence of processing, food interaction and consumer perception. *J. Cereal Sci.* **2008**, *48*, 243–257. [CrossRef]
9. Reshmi, S.; Sudha, M.; Shashirekha, M. Starch digestibility and predicted glycemic index in the bread fortified with pomelo (*Citrus maxima*) fruit segments. *Food Chem.* **2017**, *237*, 957–965. [CrossRef] [PubMed]
10. Gao, H.; Huang, Y.N.; Gao, B.; Xu, P.Y.; Inagaki, C.; Kawabata, J. α-Glucosidase inhibitory effect by the flower buds of *Tussilago farfara* L. *Food Chem.* **2008**, *106*, 1195–1201. [CrossRef]
11. Jeng, T.L.; Chiang, Y.C.; Lai, C.C.; Liao, T.C.; Lin, S.Y.; Lin, T.C.; Sung, J.M. Sweet potato leaf extract inhibits the simulated in vitro gastrointestinal digestion of native starch. *J. Food Drug Anal.* **2015**, *23*, 399–406. [CrossRef] [PubMed]
12. Barrett, A.; Ndou, T.; Hughey, C.A.; Straut, C.; Howell, A.; Dai, Z.; Kaletunc, G. Inhibition of α-amylase and glucoamylase by tannins extracted from cocoa, pomegranates, cranberries, and grapes. *J. Agric. Food Chem.* **2013**, *61*, 1477–1486. [CrossRef] [PubMed]
13. Mukherjee, P.K.; Kumar, V.; Kumar, N.S.; Heinrich, M. The Ayurvedic medicine *Clitoria ternatea*—From traditional use to scientific assessment. *J. Ethnopharmacol.* **2008**, *120*, 291–301. [CrossRef] [PubMed]
14. Talpate, K.A.; Bhosale, U.A.; Zambare, M.R.; Somani, R. Antihyperglycemic and antioxidant activity of *Clitorea ternatea* Linn. on streptozotocin-induced diabetic rats. *Ayu* **2013**, *34*, 433–439. [PubMed]
15. Parimaladevi, B.; Boominathan, R.; Mandal, S.C. Evaluation of antipyretic potential of *Clitoria ternatea* L. extract in rats. *Phytomedicine* **2004**, *11*, 323–326. [CrossRef] [PubMed]
16. Shyamkumar; Ishwar, B. Anti-inflammatory, analgesic and phytochemical studies of *Clitoria ternatea* Linn flower extract. *Int. Res. J. Pharm.* **2012**, *3*, 208–210.
17. Kamilla, L.; Mnsor, S.M.; Ramanathan, S.; Sasidharan, S. Antimicrobial activity of *Clitoria ternatea* (L.) extracts. *Pharmacologyonline* **2009**, *1*, 731–738.
18. Terahara, N.; Oda, M.; Matsui, T.; Osajima, Y.; Saito, N.; Toki, K.; Honda, T. Five new anthocyanins, ternatins A3, B4, B3, B2, and D2, from *Clitoria ternatea* flowers. *J. Nat. Prod.* **1996**, *59*, 139–144. [CrossRef] [PubMed]
19. Adisakwattana, S.; Ruengsamran, T.; Kampa, P.; Sompong, W. In vitro inhibitory effects of plant-based foods and their combinations on intestinal α-glucosidase and pancreatic α-amylase. *BMC Complement. Altern. Med.* **2012**, *12*, 110. [CrossRef] [PubMed]
20. Chayaratanasin, P.; Barbieri, M.A.; Suanpairintra, N.; Adisakwattana, S. Inhibitory effect of *Clitoria ternatea* flower petal extract on fructose-induced protein glycation and oxidation-dependent damages to albumin in vitro. *BMC Complement. Altern. Med.* **2015**, *15*, 27. [CrossRef] [PubMed]
21. Englyst, H.N.; Kingman, S.; Cummings, J. Classification and measurement of nutritionally important starch fractions. *Eur. J. Clin. Nutr.* **1992**, *46*, S33–S50. [PubMed]
22. Gularte, M.A.; Gómez, M.; Rosell, C.M. Impact of legume flours on quality and in vitro digestibility of starch and protein from gluten-free cakes. *Food Bioprocess Technol.* **2012**, *5*, 3142–3150. [CrossRef]

23. Mishra, S.; Monro, J.A. Digestibility of starch fractions in wholegrain rolled oats. *J. Cereal Sci.* **2009**, *50*, 61–66. [CrossRef]
24. Zhang, G.; Hamaker, B.R. Slowly digestible starch: Concept, mechanism, and proposed extended glycemic index. *Crit. Rev. Food Sci. Nutr.* **2009**, *49*, 852–867. [CrossRef] [PubMed]
25. Englyst, K.N.; Englyst, H.N.; Hudson, G.J.; Cole, T.J.; Cummings, J.H. Rapidly available glucose in foods: An in vitro measurement that reflects the glycemic response. *Am. J. Clin. Nutr.* **1999**, *69*, 448–454. [CrossRef] [PubMed]
26. Suganya, G.; Sampath Kumar, P.; Dheeba, B.; Sivakumar, R. In vitro antidiabetic, antioxidant and anti-inflammatory activity of *Clitoria ternatea* L. *Int. J. Pharm. Pharm. Sci.* **2014**, *6*, 342–347.
27. Oboh, G.; Ademosun, A.O.; Ayeni, P.O.; Omojokun, O.S.; Bello, F. Comparative effect of quercetin and rutin on α-amylase, α-glucosidase, and some pro-oxidant-induced lipid peroxidation in rat pancreas. *Comp. Clin. Pathol.* **2015**, *24*, 1103–1110. [CrossRef]
28. Sheng, Z.; Ai, B.; Zheng, L.; Zheng, X.; Xu, Z.; Shen, Y.; Jin, Z. Inhibitory activities of kaempferol, galangin, carnosic acid and polydatin against glycation and α-amylase and α-glucosidase enzymes. *Int. J. Food Sci. Technol.* **2018**, *53*, 755–766. [CrossRef]
29. Homoki, J.R.; Nemes, A.; Fazekas, E.; Gyémánt, G.; Balogh, P.; Gál, F.; Al-Asri, J.; Mortier, J.; Wolber, G.; Babinszky, L.; et al. Anthocyanin composition, antioxidant efficiency, and α-amylase inhibitor activity of different Hungarian sour cherry varieties (*Prunus cerasus* L.). *Food Chem.* **2016**, *194*, 222–229. [CrossRef] [PubMed]
30. Everette, J.D.; Walker, R.B.; Islam, S. Inhibitory activity of naturally occurring compounds towards rat intestinal α-glucosidase using p-nitrophenyl-α-d-glucopyranoside (PNP-G) as a substrate. *Am. J. Food Technol.* **2013**, *8*, 65–73.
31. Podsedek, A.; Majewska, I.; Kucharska, A.Z. Inhibitory potential of red cabbage against digestive enzymes linked to obesity and type 2 diabetes. *J. Agric. Food Chem.* **2017**, *65*, 7192–7199. [CrossRef] [PubMed]
32. Zhu, F. Interactions between starch and phenolic compound. *Trends Food Sci. Technol.* **2015**, *43*, 129–143. [CrossRef]
33. Camelo-Méndez, G.A.; Agama-Acevedo, E.; Sanchez-Rivera, M.M.; Bello-Pérez, L.A. Effect on in vitro starch digestibility of Mexican blue maize anthocyanins. *Food Chem.* **2016**, *211*, 281–284. [CrossRef] [PubMed]
34. Wu, Y.; Lin, Q.; Chen, Z.; Xiao, H. The interaction between tea polyphenols and rice starch during gelatinization. *Food Sci. Technol. Int.* **2011**, *17*, 569–577. [CrossRef] [PubMed]
35. Englyst, K.N.; Liu, S.; Englyst, H.N. Nutritional characterization and measurement of dietary carbohydrates. *Eur. J. Clin. Nutr.* **2007**, *61*, S19–S39. [CrossRef] [PubMed]
36. Calixto, F.S.; Abia, R. Resistant starch: An indigestible fraction of foods. *Grasas Aceites* **1991**, *42*, 239–242. [CrossRef]
37. Cone, J.W.; Wolters, M.G. Some properties and degradability of isolated starch granules. *Starch-Stärke* **1990**, *42*, 298–301. [CrossRef]
38. Alcázar-Alay, S.C.; Almeida-Meireles, M.A. Physicochemical properties, modifications and applications of starches from different botanical sources. *Food Sci. Technol.* **2015**, *35*, 215–236. [CrossRef]
39. Meynier, A.; Goux, A.; Atkinson, F.; Brack, O.; Vinoy, S. Postprandial glycaemic response: How is it influenced by characteristics of cereal products? *Br. J. Nutr.* **2015**, *113*, 1931–1939. [CrossRef] [PubMed]
40. Slavin, J.L. Carbohydrates, dietary fiber, and resistant starch in white vegetables: Links to health outcomes. *Adv. Nutr.* **2013**, *4*, 351S–355S. [CrossRef] [PubMed]
41. Ells, L.J.; Seal, C.J.; Kettlitz, B.; Bal, W.; Mathers, J.C. Postprandial glycaemic, lipaemic and haemostatic responses to ingestion of rapidly and slowly digested starches in healthy young women. *Br. J. Nutr.* **2005**, *94*, 948–955. [CrossRef] [PubMed]
42. Keeratipibul, S.; Luangsakul, N.; Lertsatchayarn, T. The effect of Thai glutinous rice cultivars, grain length and cultivating locations on the quality of rice cracker (arare). *LWT-Food Sci. Technol.* **2008**, *41*, 1934–1943. [CrossRef]
43. Hu, P.; Zhao, H.; Duan, Z.; Linlin, Z.; Wu, D. Starch digestibility and the estimated glycemic score of different types of rice differing in amylose contents. *J. Cereal Sci.* **2004**, *40*, 231–237. [CrossRef]
44. Chan, H.; Brand-Miller, J.; Holt, S.; Wilson, D. The glycaemic index values of Vietnamese foods. *Eur. J. Clin. Nutr.* **2001**, *55*, 1076–1083. [CrossRef] [PubMed]

45. Ranawana, D.; Henry, C.; Lightowler, H.; Wang, D. Glycaemic index of some commercially available rice and rice products in Great Britain. *Int. J. Food Sci. Nutr.* **2009**, *60*, 99–110. [CrossRef] [PubMed]

46. Lemlioglu-Austin, D.; Turner, N.D.; McDonough, C.M.; Rooney, L.W. Effects of sorghum [*Sorghum bicolor* (L.) Moench] crude extracts on starch digestibility, estimated glycemic index (EGI), and resistant starch (RS) contents of porridges. *Molecules* **2012**, *17*, 11124–11138. [CrossRef] [PubMed]

47. Dupuis, J.H.; Lu, Z.H.; Yada, R.Y.; Liu, Q. The effect of thermal processing and storage on the physicochemical properties and in vitro digestibility of potatoes. *In. J. Food Sci. Technol.* **2016**, *51*, 2233–2241. [CrossRef]

48. Goh, R.; Gao, J.; Ananingsih, V.K.; Ranawana, V.; Henry, C.J.; Zhou, W. Green tea catechins reduced the glycaemic potential of bread: An in vitro digestibility study. *Food Chem.* **2015**, *180*, 203–210. [CrossRef] [PubMed]

49. Coe, S.; Fraser, A.; Ryan, L. Polyphenol bioaccessibility and sugar reducing capacity of black, green, and white teas. *Int. J. Food Sci.* **2013**, *2013*, 1–6. [CrossRef] [PubMed]

50. Sivam, A.S.; Sun-Waterhouse, D.; Quek, S.; Perera, C.O. Properties of bread dough with added fiber polysaccharides and phenolic antioxidants: A review. *J. Food Sci.* **2010**, *75*, R163–R174. [CrossRef] [PubMed]

51. Han, H.M.; Koh, B.K. Effect of phenolic acids on the rheological properties and proteins of hard wheat flour dough and bread. *J. Sci. Food Agric.* **2011**, *91*, 2495–2499. [CrossRef] [PubMed]

52. Pathak, D.; Majumdar, J.; Raychaudhuri, U.; Chakraborty, R. Study on enrichment of whole wheat bread quality with the incorporation of tropical fruit by-product. *Int. Food Res. J.* **2017**, *24*, 238–246.

53. Huang, G.; Guo, Q.; Wang, C.; Ding, H.H.; Cui, S.W. Fenugreek fibre in bread: Effects on dough development and bread quality. *LWT-Food Sci. Technol.* **2016**, *71*, 274–280. [CrossRef]

Article

Buchanania obovata: Functionality and Phytochemical Profiling of the Australian Native Green Plum

Selina A. Fyfe [1,2], **Gabriele Netzel** [2], **Michael E. Netzel** [2] and **Yasmina Sultanbawa** [2,*]

[1] School of Agriculture and Food Sciences, The University of Queensland, St Lucia, Brisbane, QLD 4072 Australia; selina.fyfe@uqconnect.edu.au

[2] Queensland Alliance for Agriculture and Food Innovation (QAAFI), The University of Queensland, Health and Food Sciences Precinct, 39 Kessels Rd Coopers Plain;, PO Box 156, Archerfield, QLD 4108, Australia; g.netzel@uq.edu.au (G.N.); m.netzel@uq.edu.au (M.E.N.)

* Correspondence: y.sultanbawa@uq.edu.au; Tel.: +61-7-3443-2471

Received: 13 April 2018; Accepted: 26 April 2018; Published: 4 May 2018

Abstract: The green plum is the fruit of *Buchanania obovata* Engl. and is an Australian Indigenous bush food. Very little study has been done on the green plum, so this is an initial screening study of the functional properties and phytochemical profile found in the flesh and seed. The flesh was shown to have antimicrobial properties effective against gram negative (*Escherichia coli* 9001—NCTC) and gram positive (*Staphylococcus aureus* 6571—NCTC) bacteria. Scanning electron microscopy analysis shows that the antimicrobial activity causes cell wall disintegration and cytoplasmic leakage in both bacteria. Antioxidant 2,2-diphenyl-1-picrylhydrazyl (DPPH) testing shows the flesh has high radical scavenging activity (106.3 ± 28.6 µM Trolox equivalant/g Dry Weight in methanol). The flesh and seed contain a range of polyphenols including gallic acid, ellagic acid, p-coumaric acid, kaempferol, quercetin and trans-ferulic acid that may be responsible for this activity. The seed is eaten as a bush food and contains a delphinidin-based anthocyanin. The green plum has potential as a functional ingredient in food products for its antimicrobial and antioxidant activity, and further investigation into its bioactivity, chemical composition and potential applications in different food products is warranted.

Keywords: *Buchanania obovata*; green plum; fruit; polyphenols; antioxidants; Indigenous Australia

---

1. Introduction

Buchanania obovata Engl. is a native Australian tree that grows in the northern parts of Australia in Western Australia and the Northern Territory [1]. It produces a small green fruit known as the green plum which is eaten by Indigenous Australians. As a food it is eaten straight from the tree or with the flesh and seeds mashed into a paste, and is a favourite with children [2,3].

The *B. obovata* plant is in the family Anacardiaceae, which also contains the mango (*Mangifera indica*), cashew apple (*Anacardium occidentale*) and pistachio nut (*Pistacia vera*) [4].

The green plum was selected to be in this study because it is commonly eaten as a food and parts of the tree are used as bush medicine by Australian Aboriginal people. Leaf ribs, young stems, and the inner bark from young branches and older stems are used as bush medicine for their antiseptic and analgesic qualities to treat toothache, skin conditions and infections, and as an eye lotion [5,6].

Other native Australian foods have good nutritional and functional properties. The Kakadu plum (*Terminalia ferdinandiana*) has very high levels of ascorbic acid, high antioxidant capacity [7,8], and strong antimicrobial activity, enabling it to be used as a natural preservative for the commercial dipping of prawns in Queensland, Australia [9]. It is increasingly being used for its functional properties and is currently used in food and beauty products [10]. Davidsons plum (*Davidsonia pruriens* F. Muelli) has shown vitro anti-proliferative activity against a variety of cancer cells [11] and can be used as a natural preservative in meat [9].

Synthetic antioxidants like butylated hydroxytoluene (BHT), butylated hydroxyanisole (BHA) and propyl gallate are commonly used in food products, but have been associated with carcinogenic and toxic effects [12]. Risk perceptions of chemicals in foods cause people to prefer natural foods [13]. A total of 84% of people in the 2017 Eurobarometer survey said they are worried about the impact of chemicals present in everyday products on their health [14]. The food industry is increasingly trying to replace synthetic antioxidants with natural ingredients that are safer. There is interest in the use of foods that inherently contain bioactive compounds with these properties and can be used as food additives and ingredients [15].

This study seeks to determine the functional properties (antimicrobial and antioxidant) of the green plum flesh and seed and the components that give it these properties. This is the first study of this kind on the green plum and will give an understanding for its potential use as both a food and as a functional ingredient in food processing.

2. Materials and Methods

2.1. Chemicals

The chemicals used were from Sigma-Aldrich (Castle Hill, NSW, Australia). The Standard Plate Count Agar (APHA) CM0463 was from Oxoid Ltd., (Basingstroke, England).

2.2. Sample Preparation

The *B. obovata* fruit were hand-picked fresh at a maturity stage that was slightly under-ripe to optimise the phytochemicals and for ease of transportation. They were picked from trees near Darwin, Northern Territory, Australia, in 2015. The whole fruit were frozen, transported to Brisbane and remained frozen at -20 °C. A total of 554 g of the whole fruit with seeds intact (approximately 700 green plums) were lyophilised in a Christ Gamma 1-16 LSC Freeze Drying Unit (John Morris Scientific, Osterode, Germany), then flesh and seed were separated by blending the flesh to a powder in a kitchen food processor. The seeds were milled into a powder using a hammer mill (Lab Mill, Christy and Norris Ltd., Chelmsford, England). The composite flesh and seed powders were stored separately at room temperature, protected from light and in air-tight plastic containers.

2.3. Accelerated Solvent Extractions

Extractions were carried out using an Accelerated Solvent Extractor (ASE) (ASE 350, Dionex Corporation, Sunnyvale, CA, USA) with a method slightly modified from that in Navarro et al. [16]. Briefly, extractions were performed on 1 g samples of flesh powder or seed powder in triplicate using 100% methanol, 95% ethanol or distilled water as solvents, at 60 °C, 60 °C and 80 °C respectively with eight cycles. The extractions were filtered then evaporated until dry in a miVac evaporator (Genevac Ltd., Ipswich, England). The resulting extract powder was stored in air tight plastic containers at -20 °C.

2.4. Extraction of Unbound Polyphenols

The unbound polyphenol extraction method was slightly modified from Kammerer et al. [17]. Samples of 200 mg of flesh and seed powder were extracted in triplicate with methanol:water: hydrochloric acid (80:19:1) three times on a reciprocating shaker (RP1812, Paton Scientific, Stepney, SA, Australia) at 200 rpm for 2 h at ambient temperature, then centrifuged and the supernatant was removed. The supernatant and residue were stored separately at -20 °C.

2.5. Extraction of Bound Polyphenols

The extraction of the bound polyphenols was modified from Adom and Liu [18]. The residue from the unbound polyphenol extraction was hydrolysed for 1 h at 200 rpm in 2 M NaOH on a reciprocating shaker at ambient temperature. It was acidified to pH 2.0 with concentrated HCl and extracted five times with ethyl acetate. The ethyl acetate layers were dried under nitrogen at 40 °C (Dry Block Heater, Ratek Instruments, Boronia, VIC, Australia) before being reconstituted in 50% methanol and stored at −20 °C.

2.6. In Vitro Antioxidant Capacity

ASE extraction aliquots were made up in their extraction solvent to 1 mg/mL with dilutions, and tests were done in triplicate from the triplicate extracts of each solvent ($n = 9$). Tests were done in 96-well plates and read on a Tecan Microplate Reader (Tecan Infinite M200, Tecan Trading AG, Mannedorf, Switzerland) with Magellan Software (version 6.4, Tecan Trading AG, Mannedorf, Switzerland).

The total phenolics content (TPC) measured the reducing capacity of the flesh and seeds and was modified from Singleton and Rossi [19] and Ahmed et al. [20] with gallic acid standards on all extractions. Results are presented as g gallic acid equivalents (GAE)/kg dry weight (DW).

Radical scavenging activity was measured using a 2,2-diphenyl-1-picrylhydrazyl (DPPH) method modified from Yu and Moore [21], with Trolox standards at a concentration range of 5–35 µM/L. Briefly, equal amounts of control, standard or samples and 0.15 mM DPPH were mixed, incubated, and the absorbance read at 517 nm. Results were calculated to µM Trolox equivalents (TE)/g DW.

Chelating activity was measured using a ferrous ion chelating (FIC) assay modified from Decker and Welch [22], Kuda et al. [23] and Wang et al. [24], with ethylenediaminetetraacetic acid (EDTA) as a positive control. Briefly, 200 µL of solvent controls or samples and 10 µL of 1 mM ferrous chloride were mixed and the absorbance read at 562 nm to obtain blank results. 15 µL of 2.5 mM ferrozine was added, it was incubated (10 min, dark, ambient temperature), and the absorbance read at 562 nm. The blank was subtracted from the result reading and % chelating was calculated using the solvent control.

2.7. Antimicrobial Activity

Antimicrobial activity was tested on each type of ASE extraction using a Kirby–Bauer disc diffusion assay modified from Dussault et al. [25]. Standard Plate Count Agar (APHA CM0463, Oxoid Ltd., Hampshire, England) was spread with either *Staphyloccocus aureus* strain 6571 (NCTC—National Collection of Type Cultures, Health Protection Agency Centre for Infection, London, UK) or *Escherichia coli* strain 9001 (NCTC) at a McFarland turbidity of absorbance 0.1 at 540 nm. Triplicates of 75 µL of concentrated extract in 20% ethanol were added to aseptically placed discs on the bacteria coated agar, with 20% ethanol as the control. The concentration of extract on each disc was equivalent to approximately 130 mg DW of flesh freeze dried powder and 60 mg DW of seed freeze dried powder. Plates were incubated (Sanyo Incubator, MIR-154, Sanyo Electric Co., Ltd., Osaka, Japan) at 37 °C for 24 h. The diameter of the zone of inhibition was measured in mm under a lighted magnifying glass using a 150 mm digital caliper (Craftright Engineering Works, Jiangsu, China).

2.8. Scanning Electron Microscopy of Antimicrobial Activity

The same *S. aureus* and *E. coli* strains were grown for 7 h in tryptone soy yeast extract broth (TSYEB) at 37 °C. A total of 75 µL of concentrated green plum flesh water extract in 20% ethanol (from approximately 130 mg DW of flesh powder) was added to 1 mL bacteria and broth samples, and 75 µL of 20% ethanol was added to the controls, and incubated for 24 h at 37 °C. The samples and controls were washed three times in sterile phosphate buffered saline and fixed in 3% glutaraldehyde. They were adhered to poly-L-lysine-coated (1 mg/mL) coverslips and dehydrated in ethanol before being dried in a critical point dryer (Tousimis Research Corporation, Rockville, MD, USA). Coverslips were attached to stubs with double-sided carbon tabs and coated with gold before samples were imaged using scanning electron microscopy (SEM) on a Jeol Neoscope JCM 5000 (Jeol Ltd., Tokyo, Japan) at an accelerating voltage of 10 kV.

2.9. Phytochemical Quantification and Identification

The phytochemicals in the extracts of the free and bound polyphenols as well as the ASE extracts were quantified by ultra pressure liquid chromatography—photo diode array detector (UPLC-PDA) following a modified method of Gasperotti et al. [26]. The compounds were separated on a Waters Acquity HSS T3 (100 × 2.1 mm; 1.8 µm) column at 40 °C using the following gradient: 5–20%B (3 min), isocratic (1.3 min), 20–45%B (4.7 min), and 45–100%B (2 min). Mobile phase A consisted of water and mobile phase B of acetonitrile, both containing 0.1% formic acid. The flow was set to 0.4 mL/min and 5 µL of the pre-filtered sample (0.2 µm, Pall, Cheltenhan, VIC, Australia) was injected into the system. The compounds were quantified using external calibrations of the individual phenolic compounds and the peak identity was confirmed by LC-MS.

The same method as described above was used on a ultra high pressure liquid chromatography— mass spectrometry (UHPLC-MS) QExactive (Thermo Fisher Scientific, Bremen, Germany) for peak identification. The LCMS was used in the negative mode, and were run against 39 different standards as a screening study. The seed unbound extraction had anthocyanin identified on a UHPLC-MS QExactive in the positive mode and quantified on a UHPLC with a PDA detector (Agilent, Wilmington, DE, USA).

2.10. Statistical Analysis

Data analysis and the calculations of results was carried out using Microsoft Excel software, version 2013 (Microsoft Corporation, Redmond, WA, USA) and Minitab 16 Statistical Software (Minitab Inc., State College, PA, USA). Data is presented as arithmetic means ± standard deviations.

3. Results and Discussion

3.1. In Vitro Antioxidant Capacity

The in vitro antioxidant capacity of the flesh and seed are given in Table 1. In all three tests, the methanol extractions gave higher results than the ethanol extractions, which were higher than the water extractions apart from the FIC on the seed. The percent radical scavenging activity of the green plum flesh and seed was measured at a concentration of 1 mg/mL of extract powders, and the flesh gave 93.6 ± 0.6% in methanol, 92.5 ± 0.6% in ethanol and 92.7 ± 1.0% in water. The seed had similarly high levels at the same concentration of extract powder with methanol 93.2 ± 0.3%, ethanol 92.4 ± 0.7% and water 90.8 ± 1.3%. These levels are similar to that of the Kakadu plum, and higher than the Australian native wild lime, finger lime and almost twice that of the Davidson's plum in non-acidified extracts [27]. The radical scavenging activity is much higher than the chelating ability in both the flesh and seed at the same concentration, which could affect its application in food products.

Table 1. In vitro antioxidant capacity of green plum flesh and seed freeze dried powders.

Extraction	TPC Flesh g GAE/kg DW	TPC Seed g GAE/kg DW	DPPH Flesh µM TE/g DW	DPPH Seed µM TE/g DW	FIC Flesh * % Chelating	FIC Seed * % Chelating
Methanol	19.2 ± 4.4 [b]	2.6 ± 0.6 [b]	106.3 ± 28.6 [a]	14.6 ± 3.6 [a]	21.4 ± 8.9 [a]	20.8 ± 9.4 [b]
Ethanol	10.3 ± 3.0 [c]	N/A	40.2 ± 13.4 [b]	10.0 ± 3.8 [b]	4.9 ± 6.4 [b]	17.8 ± 16.6 [b]
Water	5.0 ± 1.6 [d]	1.4 ± 0.2 [c]	34.7 ± 13.4 [b]	11.5 ± 2.8 [a,b]	3.5 ± 5.2 [b]	34.9 ± 7.3 [a]
Unbound	84.6 ± 5.3 [a]	6.5 ± 0.5 [a]	N/A	N/A	N/A	N/A
Bound	9.2 ± 2.8 [c,d]	1.1 ± 0.2 [c]	N/A	N/A	N/A	N/A

$n = 9$; * FIC % Chelating at concentration of 1 mg/mL of extract; N/A not available, mean values of each column with a different alphabet letter are significantly different (at $p < 0.05$).

The TPC of the unbound extraction was eight times higher than the Australian desert lime and lemon aspen, nearly three times higher than the quandong and riberry and one and a half times higher than both types of Davidson's plum studied by Konczak et al. [28]. It was lower than the Kakadu plum (approx. 158 g GAE/kg DW) but over twice the content of the blueberry control [28].

3.2. Antimicrobial Activity

Table 2 shows the antimicrobial results of the disc diffusion assay and Figure 1 shows the flesh results.

Table 2. Antimicrobial activity by disc diffusion assay of the green plum flesh and seed, results as diameter of inhibition zone in mm using a 6 mm disc.

Extraction	Flesh *E. coli* (mm)	Flesh *S. aureus* (mm)	Seed *E. coli* (mm)	Seed *S. aureus* (mm)
Methanol	20.73 ± 1.27 [a]	22.61 ± 1.42 [a]	NI	8.83 ± 0.24 [a]
Ethanol	9.55 ± 2.40 [c]	20.38 ± 1.14 [b]	NI	NI
Water	16.48 ± 0.95 [b]	23.44 ± 0.85 [a]	NI	6.81 ± 0.10 [b]

n = 3 All controls had no inhibition. NI: No inhibition observed. Mean values of each column with a different alphabet letter are significantly different (at $p < 0.05$).

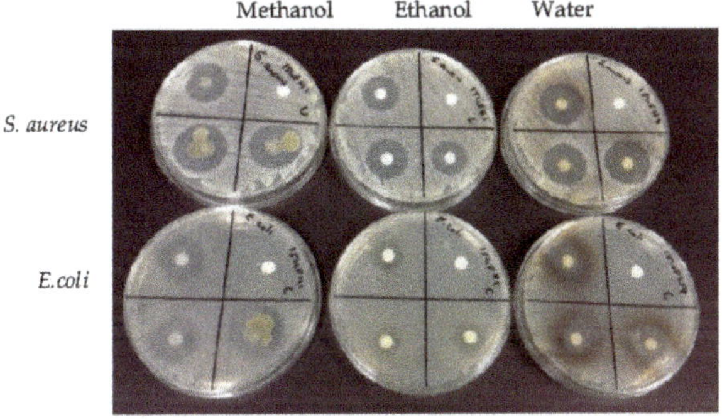

Figure 1. Disc diffusion assay of flesh against top row, *S. aureus* and bottom row *E. coli*, from left to right extracts are methanol, ethanol and water, with controls in top right hand corner of plates. n = 3.

The flesh results show that it has much higher antimicrobial activity than the seed. The flesh inhibited the growth of bacteria in all extracts for both types of bacteria. The seed only gave a low amount of inhibition for the gram positive *S. aureus* and none for the *E. coli*. Aboriginal Australians use parts of *B. obovata* plant as bush medicine for its antimicrobial properties and this work adds to the growth inhibition found by Barr et al. [6] for the leaves and twigs. Figure 1 shows that in the water extractions the flesh both inhibited growth and also appears to have turned them a brown colour.

Further investigation of the antimicrobial properties was carried out using SEM imaging on the same bacteria types and the water extraction of the flesh. Figure 2 shows the SEM images of the controls and green plum flesh water extract treatment of *S. aureus* and *E. coli*.

The *S. aureus* control cells showed smooth round surfaces and membrane integrity. The treated *S. aureus* showed shape changes, with misshapen cells, broken open cells, swelling and changes to the cell structure. There was significant leakage of cytoplasmic inclusions and some cells showed this occurring. There were cells with changes to their morphology with deep wrinkles in them, as well as indentations and changes to their size from swelling. One had clearly broken open and the cell wall opened out, and others appeared to be broken bits of cell wall. Although there were a few cells that appeared more intact, their numbers were much lower than in the control. The *E. coli* control cells generally displayed smooth surfaces with occasional surface pitting and possible cell death from 31 h of growing. In the image from the treatment there is evidence of complete cell collapse, pitting and breaking in the side wall of the cells, misshapen cells and leakage of cytoplasmic components. These indicate that there was an almost complete collapse of the cell structure, which was accompanied by cell lysis. The images show that the flesh extract in water does have antimicrobial activity. Cell disintegration may be what caused the dark colour in the flesh water extract disc diffusion plate.

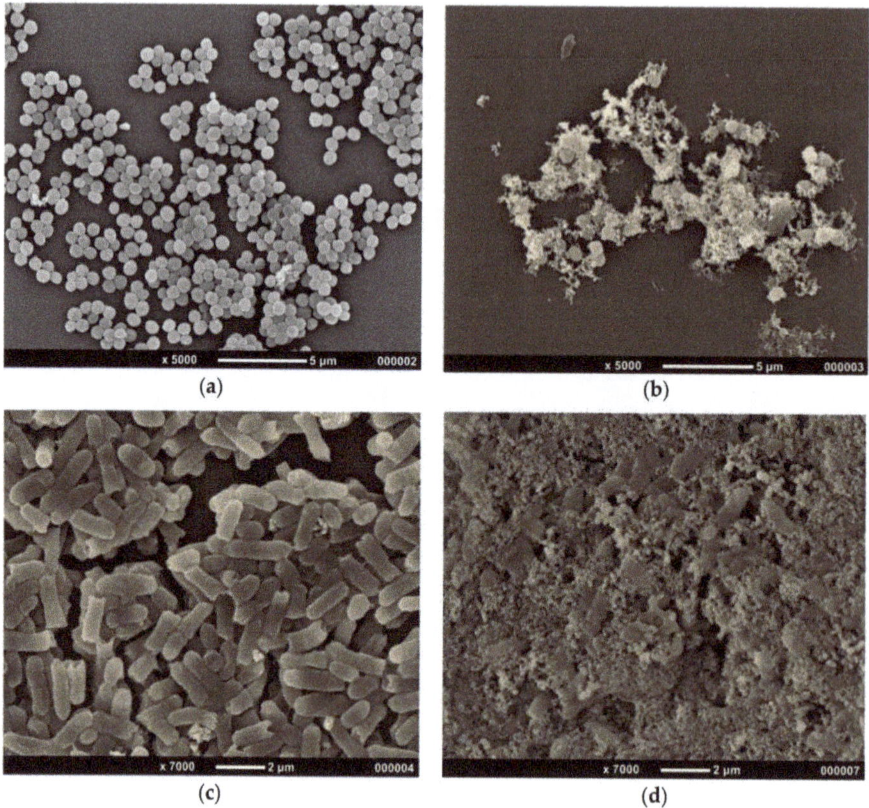

Figure 2. Scanning electron microscopy images of (**a**) *S. aureus* control; (**b**) *S. aureus* with treatment from green plum flesh water extract; (**c**) *E. coli* control and (**d**) *E. coli* with treatment from green plum flesh water extract.

These images show changes that are similar to those shown by Zhang et al. [29] in their study on the antimicrobial activity of D-limonene. The deformation and distortion caused by the green plum extract indicates strong antimicrobial activity on the cellular integrity, similar to that shown by the combined treatment of D-limonene nanoemulsions with nisin [29].

3.3. Phytochemical Analysis

The analysis of polyphenols in the green plum flesh and seed powders were carried out as a screening study of 39 different polyphenols chosen for their presence and prominence in fruit and in the mango, to which the green plum is related. The compounds identified in the various extracts are shown in Table 3. The compounds which had concentrations high enough to be quantified by UPLC-PDA are shown in Table 4.

Table 3. Compounds found in green plum flesh and seed extracts using ultra high pressure liquid chromatography—mass spectrometry (UHPLC-MS) QExactive in negative mode.

		Flesh					Seed				
	m/z	U	B	M	E	W	U	B	M	E	W
Trans-ferulic acid	193.0506	X	X	X	X	X		X	X	X	X
p-coumaric acid	163.0401	X	X	X	X	X	X		X	X	X
hydroxybenzoic acid	137.0244						X	X	X	X	X
Salicylic acid	137.0244			[x]	[x]	X					
Catechin	289.0718	X		X	X	X	X	X	X		X
Gallic acid	169.0143	X	X	X	X	X	X	X	X	X	X
Kaempferol	285.0405	X		X	X						
Naringenin	271.0612	X	X	X	X		X		X	X	
Quercetin	301.0354	X		X	X	X	X		X	X	[x]
Quercetin 3-glucoside	463.0882	X	X	X	X	X	X	X	X	X	X
Quercetin 3 rutinoside	609.1461	X	X	X	X	X	X		X	X	X
Quercetin 3-xyloside	433.0776	X	X	X	X	X			X	X	[x]
Chlorogenic acid	353.0878	X		[x]			[x]		X	X	
Cinnamic acid	147.0452	X									
Vanillic acid	167.0350						X				
Isorhamnetin-3-glucoside	477.1039	[x]		[x]	[x]	[x]	X		X	X	
Quercetin 3,4'-diglucoside	625.1410	X		X	X	X	X		[x]	X	X
Procyanidin B1	577.1352	[x]		[x]	[x]	[x]			X		[x]
Eriodicyol 7-glucoside	449.1089	X		X	X	X			X	X	X
Ellagic acid	300.9990		X	X	X	X	X	X	X	X	X

X, compounds present; [x], compound present in trace amounts; U, unbound; B, bound; M, methanol; E, ethanol; W, water. *n* = 1.

Table 4. Quantified phytochemicals in green plum flesh and seed unbound and bound extractions.

	Flesh Unbound (µg/g DW)	Flesh Bound (µg/g DW)	Seed Unbound (µg/g DW)	Seed Bound (µg/g DW)
Gallic Acid	955.39	3342.97	151.89	3.95
Myricetin	180.96			
Quercetin 3,4'-diglucoside				63.37
Chlorogenic acid	91.83	19.93	695.19	45.71
Ferulic Acid	114.50	8.02	22.77	17.85
p-coumaric acid			4.42	
Quercetin 3-glucoside	874.17	39.64	13.19	
Quercetin	118.59			

The green plum flesh and seed contain a number of compounds with antioxidant and antimicrobial activity as well as other beneficial properties. The high radical scavenging ability shown in the DPPH assay may be due to the presence of gallic acid, which was present in all extractions. Gallic acid is able to scavenge a variety of free radicals including singlet oxygen, hydroxyl, peroxyl and alkyl peroxyl and protects against UV cell damage [30]. *Trans*-ferulic acid is a low toxic component of plants and protects against pathogen invasion and has antioxidant, antimicrobial and anti-inflammatory activities [31]. It scavenges reactive oxygen species and reactive nitrogen species and nanoparticles containing it reduce lipid peroxidation [32].

Chlorogenic acid inhibits a range of bacterial pathogens by increasing outer membrane permeability, inducing the efflux of potassium from the cell, and causing rupture of the membrane and leakage of the cytoplasmic contents, including nucleotides [33].

Some flavonoid combinations can work synergistically in their antioxidant capacity, including kaempferol and quercetin (+19.9%) and catechin and kaempferol (+2.4%) [34], all found in the green plum flesh. Catechin and quercetin may mitigate adipose inflammation, oxidative stress and insulin resistance and have protective effects on human health [35]. They are antioxidants with radical scavenging activity and reducing properties [36] and quercetin is active against oxidation deterioration in emulsions [15].

Naringenin was also found and it has antioxidant, anti-carcinogenic and anti-inflammatory effects [37]. Ellagic acid has antioxidant functions and has an inhibitory effect on the growth of micro-organisms [38]. There is an abundance of ellagic acid in the Kakadu plum that contributes to its use as a natural food preservative [7], and it was found in both the flesh and seed of the green plum.

The unbound seed extractions were pink in colour and was analysed for anthocyanin. The spectrum was characteristic for anthocyanin compounds and the associated fragments in the LC-MS showed an m/z of 303.0482 and a structure of $C_{15}H_{11}O_7$, which is indicative of a delphinidin aglycone. The quantification was done against a cyanidin 3-glucoside calibration curve and the concentration of the anthocyanin was 423 ± 8.04 µg cyanidin 3-glucoside equivalents/g DW in the seed. Anthocyanins have been found previously in the seed coats of common beans using a similar extraction method [39], and in the seed coats of black soybeans, including delphinidin-3-glucoside [40].

Antioxidants are important in food to prevent the oxidation that occurs naturally and causes off-flavours and rancidity, loss of fat-soluble vitamins, fatty acids and bioactives, and can sometimes form toxic compounds [41]. There is an increased interest from consumers and the food industry in replacing synthetic preservatives with natural ones, such as plant material that inherently has these properties [15]. Other Australian native plant foods are being incorporated into food products for their natural preservative effects [9]. Plant extracts can be used in the food industry to increase the shelf-life of food products, using their natural antioxidant and antimicrobial properties to control the growth of food-borne pathogens [42]. The green plum could potentially be used as a natural alternative to chemical additives to increase the shelf life and quality of food. The antioxidant and antimicrobial potential of the green plum indicate it may be able to be used as a functional ingredient.

4. Conclusions

The green plum has considerable antimicrobial activity in its flesh, the water extraction caused bacterial cell deformation and disintegration. The radical scavenging activity is high particularly in the flesh. These activities may be the result of a number of the phytochemicals found in its flesh and seed including gallic acid, ellagic acid, p-coumaric acid, kaempferol, trans-ferulic acid and quercetin. The seed was also found to contain anthocyanins that have yet to be identified completely. The green plum is an Australian native fruit that has potential for use as a food preservative and further investigation into these uses is justified.

Author Contributions: Y.S. conceived and designed the experiments; S.F. performed the extractions and experiments; M.N. designed the phytochemical analysis; G.N. performed the phytochemical analysis; S.F. and G.N. analyzed the data; Y.S. contributed reagents/materials/analysis tools; S.F. wrote the paper and Y.S., G.N. and M.N. checked and edited it.

Acknowledgments: We thank Chris Brady and the Thamarrurr Rangers from the Northern Territory for assisting in collecting the green plum fruit, and Kathryn Green at the Centre for Microscopy and Microanalysis, The University of Queensland, for taking the SEM images. Funding for this project was from The University of Queensland, Brisbane, Australia. Selina Fyfe's Ph.D. is supported by an Australian Government Research Training Program Scholarship.

References

1. McMahon, G. *Bushtucker in the Top End*; Northern Territory Government: Northern Territory, Australia, 2005.
2. Head, L. Country and garden: Ethnobotany, archaeobotany and Aboriginal landscapes near the Keep River, Northwestern Australia. *J. Soc. Archaeol.* **2002**, *2*, 173–196. [CrossRef]
3. Medley, P.; Bollhöfer, A. Influence of group II metals on Radium-226 concentration ratios in the native green plum (*Buchanania obovata*) from the Alligator Rivers Region, Northern Territory, Australia. *J. Environ. Radioact.* **2016**, *151*, 551–557. [CrossRef] [PubMed]
4. Özçelik, B.; Aslan, M.; Orhan, I.; Karaoglu, T. Antibacterial, antifungal, and antiviral activities of the lipophylic extracts of *Pistacia vera*. *Microbiol. Res.* **2005**, *160*, 159–164. [CrossRef] [PubMed]
5. Levitt, D. *Plants and People: Aboriginal Uses of Plants on Groote Eylandt*; Australian Institute of Aboriginal Studies: Canberra, Australia, 1981.
6. Barr, A.; Chapman, J.; Smith, N.; Wightman, G.; Knight, T.; Mills, L.; Andrews, M.; Alexander, V. *Traditional Aboriginal Medicines in the Northern Territory of Australia, by Aboriginal Communities of the Northern Territories*; Conservation Commission of the Northern Territory of Australia: Darwin, Australia, 1993.

7. Williams, D.J.; Edwards, D.; Pun, S.; Chaliha, M.; Burren, B.; Tinggi, U.; Sultanbawa, Y. Organic acids in Kakadu plum (*Terminalia ferdinandiana*): The good (ellagic), the bad (oxalic) and the uncertain (ascorbic). *Food Res. Int.* **2016**, *89*, 237–244. [CrossRef] [PubMed]

8. Konczak, I.; Maillot, F.; Dalar, A. Phytochemical divergence in 45 accessions of *Terminalia ferdinandiana* (*Kakadu plum*). *Food Chem.* **2014**, *151*, 248–256. [CrossRef] [PubMed]

9. Sultanbawa, Y. Food Preservation and the Antimicrobial Activity of Australian Native Plants. In *Australian Native Plants: Cultivation and Uses in the Health and Food Industries*; Sultanbawa, Y., Sultanbawa, F., Eds.; CRC Press, Taylor & Francis Group: Boca Raton, FL, USA, 2016; pp. 251–263.

10. Gorman, J.; Courtenay, K.; Brady, C. Production of *Terminalia ferdinandiana* Excell. ('Kakadu Plum') in Northern Australia. In *Australian Native Plants: Cultivation and Uses in the Health and Food Industries*; Sultanbawa, Y., Sultanbawa, F., Eds.; CRC Press, Taylor & Francis Group: Boca Raton, FL, USA, 2016; pp. 89–103.

11. Chuen, T.L.K.; Vuong, Q.V.; Hirun, S.; Bowyer, M.C.; Predebon, M.J.; Goldsmith, C.D.; Sakoff, J.A.; Scarlett, C.J. Antioxidant and anti-proliferative properties of Davidson's plum (*Davidsonia pruriens* F. Muell) phenolic-enriched extracts as affected by different extraction solvents. *J. Herb. Med.* **2016**, *6*, 187–192. [CrossRef]

12. Gülçin, İ. Antioxidant activity of food constituents: An overview. *Arch. Toxicol.* **2012**, *86*, 345–391. [CrossRef] [PubMed]

13. Dickson-Spillmann, M.; Siegrist, M.; Keller, C. Attitudes toward chemicals are associated with preference for natural food. *Food Qual. Preference* **2011**, *22*, 149–156. [CrossRef]

14. European Commission. *Special Eurobarometer 468: Attitudes of European Citizens towards the Environment*; European Commission: Brussels, Belgium, 2017.

15. Kiokias, S.; Varzakas, T. Activity of flavonoids and β-carotene during the auto-oxidative deterioration of model food oil-in water emulsions. *Food Chem.* **2014**, *150*, 280–286. [CrossRef] [PubMed]

16. Navarro, M.; Sonni, F.; Chaliha, M.; Netzel, G.; Stanley, R.; Sultanbawa, Y. Physicochemical assessment and bioactive properties of condensed distillers solubles, a by-product from the sorghum bio-fuel industry. *J. Cereal Sci.* **2016**, *72*, 10–15. [CrossRef]

17. Kammerer, D.; Claus, A.; Carle, R.; Schieber, A. Polyphenol screening of pomace from red and white grape varieties (*Vitis vinifera* L.) by HPLC-DAD-MS/MS. *J. Agric. Food Chem.* **2004**, *52*, 4360–4367. [CrossRef] [PubMed]

18. Adom, K.K.; Liu, R.H. Antioxidant activity of grains. *J. Agric. Food Chem.* **2002**, *50*, 6182–6187. [CrossRef] [PubMed]

19. Singleton, V.; Rossi, J.A. Colorimetry of total phenolics with phosphomolybdic-phosphotungstic acid reagents. *Am. J. Enol. Viticult.* **1965**, *16*, 144–158.

20. Ahmed, F.; Fanning, K.; Netzel, M.; Turner, W.; Li, Y.; Schenk, P.M. Profiling of carotenoids and antioxidant capacity of microalgae from subtropical coastal and brackish waters. *Food Chem.* **2014**, *165*, 300–306. [CrossRef] [PubMed]

21. Yu, L.; Moore, J. Methods for Antioxidant Capacity Estimation of Wheat and Wheat-Based Food Products. In *Wheat Antioxidants*; Yu, L., Ed.; Wiley-Interscience: Hoboken, NJ, USA, 2008.

22. Decker, E.A.; Welch, B. Role of ferritin as a lipid oxidation catalyst in muscle food. *J. Agric. Food Chem.* **1990**, *38*, 674–677. [CrossRef]

23. Kuda, T.; Tsunekawa, M.; Hishi, T.; Araki, Y. Antioxidant properties of dried kayamo-nori', a brown alga *Scytosiphon lomentaria* (Scytosiphonales, Phaeophyceae). *Food Chem.* **2005**, *89*, 617–622. [CrossRef]

24. Wang, T.; Jonsdottir, R.; Ólafsdóttir, G. Total phenolic compounds, radical scavenging and metal chelation of extracts from Icelandic seaweeds. *Food Chem.* **2009**, *116*, 240–248. [CrossRef]

25. Dussault, D.; Vu, K.D.; Lacroix, M. In vitro evaluation of antimicrobial activities of various commercial essential oils, oleoresin and pure compounds against food pathogens and application in ham. *Meat Sci.* **2014**, *96*, 514–520. [CrossRef] [PubMed]

26. Gasperotti, M.; Masuero, D.; Mattivi, F.; Vrhovsek, U. Overall dietary polyphenol intake in a bowl of strawberries: The influence of *Fragaria* spp. in nutritional studies. *J. Funct. Foods* **2015**, *18*, 1057–1069. [CrossRef]

27. Sommano, S.; Caffin, N.; Kerven, G. Screening for Antioxidant Activity, Phenolic Content, and Flavonoids from Australian Native Food Plants. *Int. J. Food Propert.* **2013**, *16*, 1394–1406. [CrossRef]

28. Konczak, I.; Zabaras, D.; Dunstan, M.; Aguas, P.; Roulfe, P.; Pavan, A. *Health Benefits of Australian Native Foods: An Evaluation of Health-Enhancing Compounds*; Rural Industries Research and Development Corporation: Barton, Australia, 2009.

29. Zhang, Z.; Vriesekoop, F.; Yuan, Q.; Liang, H. Effects of nisin on the antimicrobial activity of D-limonene and its nanoemulsion. *Food Chem.* **2014**, *150*, 307–312. [CrossRef] [PubMed]

30. Marino, T.; Galano, A.; Russo, N. Radical scavenging ability of gallic acid toward OH and OOH radicals. Reaction mechanism and rate constants from the density functional theory. *J. Phys. Chem. B* **2014**, *118*, 10380–10389. [CrossRef] [PubMed]

31. Wang, J.; Cao, Y.; Sun, B.; Wang, C. Characterisation of inclusion complex of trans-ferulic acid and hydroxypropyl-β-cyclodextrin. *Food Chem.* **2011**, *124*, 1069–1075. [CrossRef]

32. Trombino, S.; Cassano, R.; Ferrarelli, T.; Barone, E.; Picci, N.; Mancuso, C. Trans-ferulic acid-based solid lipid nanoparticles and their antioxidant effect in rat brain microsomes. *Colloids Surf. B: Biointerfaces* **2013**, *109*, 273–279. [CrossRef] [PubMed]

33. Lou, Z.; Wang, H.; Zhu, S.; Ma, C.; Wang, Z. Antibacterial activity and mechanism of action of chlorogenic acid. *J. Food Sci.* **2011**, *76*, M398–M403. [CrossRef] [PubMed]

34. Hidalgo, M.; Sánchez-Moreno, C.; de Pascual-Teresa, S. Flavonoid–flavonoid interaction and its effect on their antioxidant activity. *Food Chem.* **2010**, *121*, 691–696. [CrossRef]

35. Vazquez Prieto, M.A.; Bettaieb, A.; Rodriguez Lanzi, C.; Soto, V.C.; Perdicaro, D.J.; Galmarini, C.R.; Haj, F.G.; Miatello, R.M.; Oteiza, P.I. Catechin and quercetin attenuate adipose inflammation in fructose-fed rats and 3T3-L1 adipocytes. *Mol. Nutr. Food Res.* **2015**, *59*, 622–633. [CrossRef] [PubMed]

36. Dueñas, M.; González-Manzano, S.; González-Paramás, A.; Santos-Buelga, C. Antioxidant evaluation of *O*-methylated metabolites of catechin, epicatechin and quercetin. *J. Pharm. Biomed. Anal.* **2010**, *51*, 443–449. [CrossRef] [PubMed]

37. Orrego-Lagarón, N.; Martínez-Huélamo, M.; Vallverdú-Queralt, A.; Lamuela-Raventos, R.M.; Escribano-Ferrer, E. High gastrointestinal permeability and local metabolism of naringenin: Influence of antibiotic treatment on absorption and metabolism. *Br. J. Nutr.* **2015**, *114*, 169–180. [CrossRef] [PubMed]

38. Landete, J. Ellagitannins, ellagic acid and their derived metabolites: A review about source, metabolism, functions and health. *Food Res. Int.* **2011**, *44*, 1150–1160. [CrossRef]

39. Díaz, A.M.; Caldas, G.V.; Blair, M.W. Concentrations of condensed tannins and anthocyanins in common bean seed coats. *Food Res. Int.* **2010**, *43*, 595–601. [CrossRef]

40. Zhang, R.F.; Zhang, F.X.; Zhang, M.W.; Wei, Z.C.; Yang, C.Y.; Zhang, Y.; Tang, X.J.; Deng, Y.Y.; Chi, J.W. Phenolic composition and antioxidant activity in seed coats of 60 Chinese black soybean (*Glycine max* L. Merr.) varieties. *J. Agric. Food Chem.* **2011**, *59*, 5935–5944. [CrossRef] [PubMed]

41. Shahidi, F.; Zhong, Y. Novel antioxidants in food quality preservation and health promotion. *Eur. J. Lipid Sci. Technol.* **2010**, *112*, 930–940. [CrossRef]

42. Sultanbawa, Y. Plant antimicrobials in food applications: Minireview. In *Science against Microbial Pathogens: Communicating Current Research and Technological Advances*; Formatex Research Center: Badajoz, Spain, 2011; pp. 1084–1099.

Article

Chemical and Nutritional Composition of *Terminalia ferdinandiana* (Kakadu Plum) Kernels: A Novel Nutrition Source

Saleha Akter [1], Michael E. Netzel [1], Mary T. Fletcher [1], Ujang Tinggi [2] and Yasmina Sultanbawa [1,*]

[1] Queensland Alliance for Agriculture and Food Innovation (QAAFI), The University of Queensland, Health and Food Sciences Precinct, 39 Kessels Rd Coopers Plains, P.O. Box 156, Archerfield BC, QLD 4108, Australia; saleha.akter@uq.edu.au (S.A.); m.netzel@uq.edu.au (M.E.N.); mary.fletcher@uq.edu.au (M.T.F.)

[2] Queensland Health Forensic and Scientific Services, Health and Food Sciences Precinct, 39 Kessels Rd, Coopers Plains, P.O. Box 594, Archerfield BC, QLD 4108, Australia; ujang.tinggi@health.qld.gov.au

* Correspondence: y.sultanbawa@uq.edu.au; Tel.: +61-734-432-471

Received: 22 March 2018; Accepted: 9 April 2018; Published: 12 April 2018

Abstract: *Terminalia ferdinandiana* (Kakadu plum) is a native Australian fruit. Industrial processing of *T. ferdinandiana* fruits into puree generates seeds as a by-product, which are generally discarded. The aim of our present study was to process the seed to separate the kernel and determine its nutritional composition. The proximate, mineral and fatty acid compositions were analysed in this study. Kernels are composed of 35% fat, while proteins account for 32% dry weight (DW). The energy content and fiber were 2065 kJ/100 g and 21.2% DW, respectively. Furthermore, the study showed that kernels were a very rich source of minerals and trace elements, such as potassium (6693 mg/kg), calcium (5385 mg/kg), iron (61 mg/kg) and zinc (60 mg/kg) DW, and had low levels of heavy metals. The fatty acid composition of the kernels consisted of omega-6 fatty acid, linoleic acid (50.2%), monounsaturated oleic acid (29.3%) and two saturated fatty acids namely palmitic acid (12.0%) and stearic acid (7.2%). The results indicate that *T. ferdinandiana* kernels have the potential to be utilized as a novel protein source for dietary purposes and non-conventional supply of linoleic, palmitic and oleic acids.

Keywords: *Terminalia ferdinandiana*; Kakadu plum; nutrition; fatty acids; proximate; minerals; kernels

---

1. Introduction

Terminalia is the second-largest genus of the combretaceae family, with approximately 250 species growing in tropical and subtropical countries around the world [1]. More than 30 species of *Terminalia* occur in northern regions of Australia [2]. More than 50 species of *Terminalia* have found utility as ingredients in foods and beverages worldwide, as preservatives, raw material for wine and palm sugar, eaten raw and as food supplements [3]. The nutritional and therapeutic properties of *Terminalia* genus can be attributed to the presence of a wide range of phytochemicals, such as phenolic compounds, which encompasses phenolic acids, gallotannins, ellagitannins, proanthocyanidins and other flavonoids [3].

Terminalia ferdinandiana, popularly known as Kakadu plum, is native to Australia. Indigenous Australians (Aboriginal people) have been using this plant as a food and medicine for centuries, for example, refreshing drinks are made from fresh or dried fruits in Western Australia [2]. Fruits are traditionally used as an antiseptic, soothing balm, in colds and flu and in treating a headache [4]. A number of research outcomes have been reported on the antioxidant [5–7], antibacterial [8,9], anti-inflammatory [10], anti-apoptotic, cytoprotective and anticancer activities [11] of *T. ferdinandiana* fruits and leaves. Phytochemical analysis has revealed that *T. ferdinandiana* fruit is a rich source of Ellagic acid and its hydrolysable tannins, ellagitannins [12]. Recently, a food safe extraction method of *T. ferdinandiana* fruits for commercial use in the food industry has been suggested [4]. Additionally, a systematic evaluation of the changes in quality and bioactivity of the fruits of *T. ferdinandiana* during processing, packaging and storage has been performed, and key chemical markers have been identified to enable standardized products to be delivered to the consumer [9].

In the last two decades, seeds and kernels from the *Terminalia* genus have been researched and reported for their nutritional properties and health-promoting activities [13–15]. To understand the relationship between the internal quality and genotype of the plant, studies of nut and kernels characteristics and composition are very common. During the industrial processing of *T. ferdinandiana* fruits, the seeds are treated as waste products and have been discarded. Recent studies on many fruit seeds or kernels have shown that they have the potential to be utilized as ingredients for value addition, they are very nutritious and could be used as alternate sources of essential minerals, fatty acids, and proteins [16–19].

To date, no reports have been published on the utilization of the by-products of *T. ferdinandiana* and there is no investigation on the chemical and nutritional composition of *T. ferdinandiana* kernels. The aim of this study was to determine the potential use of the by-product of *T. ferdinandiana* in the industry by determining proximate, mineral and fatty acid compositions to ascertain its nutritional value and potential as a source of food supplement ingredients for the food industry.

2. Materials and Methods

2.1. Sample Collection and Preparation

Fully ripe and mature fruits of *T. ferdinandiana* were collected from over 600 trees, giving a total harvest of 5000 kg, from native bush land covering an area of 20,000 km^2 in Northern Territory, Australia in 2015 and were authenticated by the experts in Queensland Herbarium, Brisbane Botanic Gardens Mt Coot-tha, Queensland, Australia, where botanical specimens were retained for future reference (AQ522453). Seeds were collected as the by-products after pureeing of the fruits, and were stored at −20 °C prior to analysis. *T. ferdinandiana* tissues are illustrated in Figure 1.

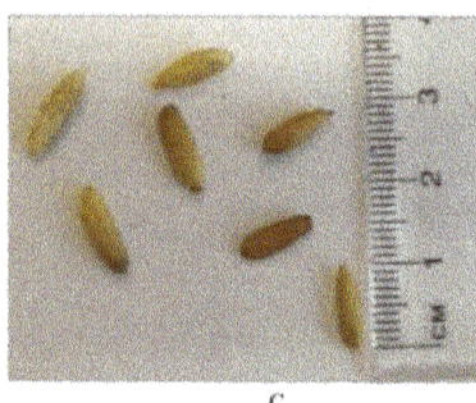

A   B   C

Figure 1. *Terminalia ferdinandiana* tissues. (**A**) Fresh fruits; (**B**) Dried seeds; (**C**) Kernels.

2.2. Processing of Seeds

The frozen seeds were thawed, washed and cleaned manually several times to remove the pulp residues with double distilled water. The seeds were then dried in the oven for 48 h at 40 °C. After drying, the seeds were individually cracked using an Engineers' vice size 125 (DAWN, Melbourne, Australia) to release the kernels from the seedcoats. The seedcoats and kernels were kept, processed and analyzed separately. The kernels were kept in air-tight containers and placed at −20 °C for further analysis. A flowchart depicting the processing of the seeds is illustrated in Figure 2. During the processing of fruits in the industry in a batch of 100 kg of fruits, 22 kg of seeds can be obtained as by-product. The average weight of a dry seed is 0.5 g and the moisture content is 2.8%. Average weight of a kernel is 0.04 g. The kernel is 8% of the weight of the seed. 1 kg of dry seeds can deliver 80 g of kernels.

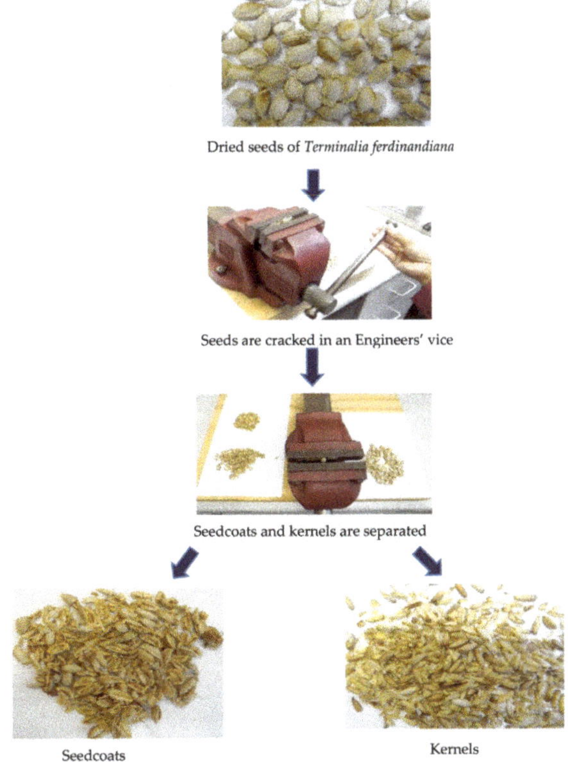

Figure 2. Processing of *Terminalia ferdinandiana* seeds to release kernels.

2.3. Proximate Composition Analysis

Physicochemical analysis of the kernels of *T. ferdinandiana* was performed at an accredited laboratory (National Association of Testing Authorities (NATA), Symbio Alliance, Eight Mile Plains, Queensland, Australia). The following analyses were done according to AOAC methods: vitamin C, protein (AOAC 990.03, 992.15 & 992.15), fat (AOAC 991.36), saturated, mono-unsaturated, polyunsaturated and trans-fat (AOAC 996.06), moisture (AOAC 925.10), ash (AOAC 923.03), sodium (using ICP-AES), total sugar (AOAC 977.20) and dietary fiber (AOAC 985.29, 991.42 and 993.19). Available carbohydrate and energy were calculated using FSANZ (Food Standards Australia New Zealand) codes.

2.4. Fatty Acid Analysis

Dried kernels (ca. 1 g) were finely chopped and extracted with chloroform and methanol (2:1) followed by agitation at room temperature for one hour. The mixture was then centrifuged for 5 min at 3500 rpm and the whole process was repeated twice. The lipid extracts were mixed with boron trifluoride (BF_3)-methanol reagent (20%) and fatty acids were derivatized to fatty acid methyl esters [20]. The methyl esters of the fatty acids were dissolved in heptane and analyzed by GC-MS (Shimadzu QP2010, Shimadzu Corporation, Tokyo, Japan). The GC conditions were as follows: Restek stabilwax capillary column (30 m × 0.25 mm ID × 0.5 µm film thickness) (Restek Corporation, Bellefonte, PA,

USA); oven temperature program: the column held initially at 100 °C after injection and the final temperature was increased to 250 °C, total program time was 39:00 min; injector temperature: 250 °C; carrier gas: Helium; linear gas velocity: 42.7 cm/s; column flow: 1.10 mL/min; split ratio: 50.00; injection volume: 1.0 μL. MS conditions were regulated as follows: ion source temperature: 200 °C; interface temperature: 250 °C; mass range: 35–500 atomic mass units. Identification of the compounds was carried out by comparison of their retention times and mass spectra with corresponding data from a standard food industry FAME Mix (Restek Corporation, Bellefonte, PA, USA). A total of 32 individual compounds were analyzed and only the detected ones were recorded along with their quantity compared with the standard.

2.5. Mineral and Trace Element Analysis

Accurately weighed 0.3 g of dried *T. ferdinandiana* kernels were taken into teflon vessels of microwave digestion system (MarsXpress, CEM, Matthews, NC, USA) and high-purity nitric acid (70% *w*/*w*, 4 mL) was added [21]. The samples were left overnight at room temperature for slow digestion gasses to evolve. The vessels were sealed and microwave-digested at increased temperature with set digestion time [22]. The digested samples were diluted and made up to 40 mL with high-purity water (Milli-Q Element system, Millipore, Bedford, MA, USA). The levels of minerals and trace elements were analyzed by inductively coupled plasma optical emission spectrometry (ICP-OES, Vista AX, Varian Australia, Mulgrave, Victoria, Australia), to measure lower levels and for greater sensitivity the analysis was carried out using ICP-MS (7500a, Agilent, Tokyo, Japan). The ICP-MS was equipped with an auto-sampler, integrated sample introduction system and a helium octopole reaction cell to remove polyatomic interferences (^{40}Ag ^{35}Cl on ^{75}As). The operating conditions were as follows: radio frequency (RF) power 1350W, argon carrier gas 0.8 L/min and helium reaction cell gas flow rate 4.5 mL/min. The standard reference materials were used for the quality control and assurance and treated similarly to the samples throughout the study. The data of quality control and assurance are presented in the Supplementary Material (Table S1).

2.6. Statistical Analysis

The data were calculated using Microsoft Excel 2013 (Microsoft Corporation, Redmond, WA, USA). The results are expressed as the mean of triplicate experiments unless otherwise specified.

3. Results and Discussion

3.1. Proximate Composition

The proximate composition of *T. ferdinandiana* kernels is summarized in Table 1. Moisture content is an important parameter in terms of the physicochemical properties of plant parts, due to the fact that low moisture content is beneficial for retaining the quality and shelf life of seeds, and this also decreases the susceptibility for microbial growth, premature seed germination, unwarranted fermentation and undesirable biochemical changes. The moisture content is only 4% in the kernels of *T. ferdinandiana*, presenting minimum risk for microbial growth and undesirable biochemical changes upon storage. A comparable moisture content of 5.5% was reported for *T. catappa* kernels [15]. Furthermore, the results of the present study showed that the kernels were abundant in proteins, with a content of 32% relative to the standard. Protein content of *T. ferdinandiana* is higher than that of *T. catappa* kernels (20.1%) [15]. *T. sericea* kernels contain 46.2% proteins [23], which is higher than *T. ferdinandiana* kernels. Recommended dietary allowances (RDA) for protein are 56 g for a 70 kg man [24]. As the protein content of *T. ferdinandiana* kernels is high, it could be used as an alternative source or dietary supplement for consumers with restricted and compromised protein intake from other sources. Ash content is 4% and dietary fiber 21.2% in *T. ferdinandiana* kernels. Ash content signifies the presence of minerals in the kernel, tissue and the high content of fiber can help in improving the gut health and digestion. The lipid content in *T. ferdinandiana* kernels was found to be 35.1%, with less than 1%

in the trans form. The WHO recommends that no more than 1% of our daily energy intake come from trans-fatty acids (TFAs). Based on the present results, it can be concluded that the fat content of *T. ferdinandiana* kernels is devoid of any trans-fat-associated health risk. The fat content in *T. sericea* seed is 32.5% [23], 64.7% in *T. catappa* kernels [15], and in *T. catappa* seed it is 32.7% [14], while in *T. ferdinandiana* kernels it is 35.1%. *T. ferdinandiana* kernels can supply 50% of the RDA of fat with saturated (SFA), monounsaturated (MUFA) and polyunsaturated (PUFA) fats are in the order of 5.8%, 9.8%, and 19.4%. These proportions are similar to the fatty acid profile determined by GC-MS (Table 2). A diet rich in PUFA is important for the structure and function of proteins, receptors, enzymes and transport molecules whereas the MUFA content may lower blood cholesterol levels, modulate immune function and can improve the fluidity of high-density lipoproteins (HDL) [25]. The results of our present study thus suggest that *T. ferdinandiana* kernels have the potential to be used as an alternative source of MUFA and PUFA.

Table 1. Proximate composition of *Terminalia ferdinandiana* kernels.

T. ferdinandiana Kernels			Nutrition Information (Servings per Package: 1) (Serving Size: 100 g)	
			Quantity per Serving	% Daily Intake */Serving
Protein	% (*w/w*)	32.0	32 g	64%
Fat	% (*w/w*)	35.1	35.1 g	50%
Saturated Fat	% (*w/w*)	5.8	5.8 g	24%
Mono-unsaturated Fat	% (*w/w*)	9.8	9.8 g	
Poly-unsaturated Fat	% (*w/w*)	19.4	19.4 g	
Trans Fat	% (*w/w*)	<0.01	<0.1 g	
Moisture (air)	% (*w/w*)	4.0		
Ash	% (*w/w*)	4.5		
Dietary Fibre (Total)	% (*w/w*)	21.2	21.2 g	71%
Dry Matter	% (*w/w*)	96.0		
Crude Fibre	% (*w/w*)	11.6		
Energy	kJ/100 g	2065	24%	
Total Sugar	g/100 g	0.49	<1 g	<1%
Available Carbohydrate	%	3.2	3.2 g	1%
Sodium (Na)	mg/100 g	8.6	8.6 mg	<1%

* Percentage daily intakes are based on an average adult diet of 8700 kJ. Results are expressed as the mean of triplicate experiments.

Table 2. Fatty acid profile of *Terminalia ferdinandiana* kernels expressed as percentage (±SD) of the total fatty acid profile as determined by FAME GC-MS analysis.

Fatty Acid	Percentage (%) ± SD
Saturated	
C14:0 Methyl myristate	0.1 ± 0.01
C16:0 Methyl palmitate	12 ± 0.53
C18:0 Methyl stearate	7.2 ± 0.13
C20:0 Methyl arachidate	0.7 ± 0.06
C22:0 Methyl behenate	0.4 ± 0.13
TSFA	**20.4**
Monounsaturated	
C16:1 (*cis*-9) Methyl palmitoleate	0.1 ± 0.06
C18:1 (*cis*-9) Methyl oleate	29.2 ± 0.68
C20:1 (*cis*-11) Methyl eicosenoate	0.1 ± 0.04
TMUFA	**29.4**
Polyunsaturated	
C18:2 (all-*cis*-9,12) Methyl linoleate	50.2 ± 1.1
PUFA	**50.2**
SFA vs. UFA	0.25:1
MUFA vs. PUFA	0.6:1

SFA: saturated fatty acids; UFA: unsaturated fatty acids; MUFA: monounsaturated fatty acids; PUFA: polyunsaturated fatty acids; TSFA: total saturated fatty acids; TMUFA: total monounsaturated fatty acids; TPUFA: total polyunsaturated fatty acids; data presented as a mean ± SD of triplicate experiments.

3.2. Mineral and Trace Element Composition

The macro and trace element composition of *T. ferdinandiana* kernels evaluated in this study is presented in Table 3, and the non-essential and heavy metal composition is presented in Table 4. Minerals are essential for proper functioning of the body, and a deviation from the appropriate amounts can cause numerous diseases, clinical syndromes, and illnesses associated with the deficient intake, as well as overuse over time or at a certain time period of life. Hence, reference values are established and reviewed periodically to stipulate the mineral levels that will meet the needs of healthy human individuals. The RDA of the evaluated minerals for a healthy male adult of 70 kg body weight are also presented in Tables 3 and 4. The high macro-mineral contents were found to be phosphorus 872.8 mg/100 g DW, and calcium at 538.5 mg/100 g DW, while sodium was 120.3 mg/100 g DW and magnesium 421.1 mg/100 g DW (Table 3). These results indicated that the kernels could significantly contribute to the mineral intake in humans. Mineral composition analysis of kernels from *Terminalia* genus is scarce and one report on the mineral composition of *T. catappa* seeds included phosphorus (10), calcium (36.1), magnesium (26.4), iron (375), sodium (5) and potassium (350), in mg/100 g [14]. Kernels from bayberry (*Myrica rubra*) were reported as an abundant source of potassium, containing 780 mg/100 g [26]. The potassium content of white Chinese olive (*Canarium album*) is also high, at 587 mg/100 g [19]. In our study, *T. ferdinandiana* kernels contained 669.3 mg/100 g of potassium, which can be compared to the potassium content of bayberry, Chinese olive, and black Chinese olive. Moreover, the phosphorus levels of *T. ferdinandiana* kernels (872.8 mg/100 g DW) seemed to be much higher compared to the levels of bayberry (32.9 mg/100 g) [16]. Important trace elements found in *T. ferdinandiana* were zinc, manganese, copper and iron at levels of 6, 9.1, 2.5 and 6.1 mg/100 g, respectively, and are within the RDA and AI values. It can be suggested that *T. ferdinandiana* kernels can be a valuable dietary source of these trace elements. These trace elements are important constituents of various proteins and enzymes of our body which are involved in macronutrient metabolism [15]. The levels of molybdenum, arsenic, mercury, cadmium were found at less than 0.1 mg/kg in the kernels, while the lead level was found at 0.13 mg/kg (Table 4). Heavy metal exposure poses significant health risks, which can cause life-threatening diseases, and the toxic effects are influenced by chemical forms, absorption rate, and solubility in body fluids. The toxicity of arsenic depends on the chemical form. The inorganic form of arsenic is more toxic than organic arsenic [21]. Mercury can be readily absorbed and incorporated into tissue proteins and can cause detrimental effects on health. The bioaccessibility and bioavailability of the exposed heavy metals can again vary depending on the chemical forms, time and route of exposure, duration, and concentration of the exposed metals. However, the levels of heavy metals found in *T. ferdinandiana* kernels were within the regulatory limits, suggesting that they may not impose any health risk.

Table 3. Major and trace elements composition of *Terminalia ferdinandiana* kernels (mg/100 g DW).

	Mineral Composition														
	Major Elements								Micro/Trace Elements						
	Ca	Mg	Na	K	P	Fe	Zn	Mn	Cu	Co	Ni	Mo	Se	Sr	B
Kernels (mg/100 g DW)	538.5	421.1	120.3	669.3	872.8	6.1	6.0	9.1	2.5	0.02	0.17	<0.01	0.02	5.0	0.83
DRI	1200 AI [a]	350 EAR [a]	1.3 AI [a]	4.7 AI [a]	700 RDA [a]	8 RDA [a]	11 RDA [a]	2.3 AI [a]	700 RDA [a]	0.12AI [a]	1.0 UL [a]	34 EAR [a]	45 EAR [a]	1–5 RDA [a]	20 UL [a]
Units	mg	mg	g	g	mg	mg	mg	mg	µg	µg	mg	µg	µg	mg	mg

Results are expressed as the mean of duplicate experiments. RDA—recommended dietary allowance, AI—adequate Intake, UL—tolerable upper intake level, DRI—Dietary reference intakes; EAR—Estimated average requirement. [a] Institute of Medicine. 2006. *Dietary Reference Intakes: The Essential Guide to Nutrient Requirements*. Washington, DC: The national academic press. ISBN: 978-0-309-15742-1. doi:10.17226/11537. URL https://www.nap.edu/read/11537/chapter/1. Accessed on 17 February 2017.

Table 4. Non-essential elements and heavy metal compositions of *Terminalia ferdinandiana* kernels (mg/100 g DW).

	Non-Essential Elements	Heavy Metals			
	Ba	As	Hg	Pb	Cd
Kernels (mg/100 g DW)	0.31	<0.01	<0.01	0.013	<0.01
DRI	0.02 UL [a]	12.5–25 UL [b]	5 UL [c]	25 UL [c]	2.5 UL [d]
Units	mg/kg BW	µg/kg BW/week	µg/kg BW/week	µg/kg BW/week	µg/kg BW/week

Results are expressed as the mean of duplicate experiments. UL—tolerable upper intake level, DRI—Dietary reference intakes; BW—body weight; [a] Scientific Committee on Health and Environmental Risks (2012). Assessment of the tolerable daily intake of barium. European commission. URL http://ec.europa.eu/health/scientific_committees/environmental_risks/docs/scher_o_161.pdf. Accessed 17 February 2017; [b] Food Safety authority of Ireland (2009). Mercury, Lead, Cadmium, Tin, and Arsenic in Food. Toxicology factsheet series, Issue no. 1. URL www.fsai.ie/WorkArea/DownloadAsset.aspx?id=8412. Accessed 17 February 2017; [c] [27]; [d] Statements on the tolerable weekly intake for cadmium. Panel on contaminants in the food chain. EFSA, 2011, 9(2). URL https://www.efsa.europa.eu/en/efsajournal/pub/1975. Accessed on 17 February 2017.

3.3. Fatty Acid Composition

Fatty acids can be considered the main constituent of all oils and may include saturated (SFA), monounsaturated (MUFA) and polyunsaturated (PUFA) fatty acids [25]. Besides providing high-quality food, vegetable oils can also provide essential nutrients that have a particular clinical significance. PUFA are present in membrane phospholipids in some tissues and can also act as precursors for prostaglandin hormones [28]. On the other hand, SFA are reported to increase cardiovascular disease risk and sometimes can potentiate the risk of cancer and autoimmune disorders [29]. The principal fatty acid components in *T. ferdinandiana* kernels were palmitic (SFA, 12%); oleic (MUFA, 29.3%); and linoleic (PUFA, 50.2%) acids (Table 2).

SFA are reported to impact human health by increasing the plasma low-density lipoprotein (LDL) cholesterol. However, some of the SFA are also reported to increase the high-density lipoprotein (HDL) cholesterol and some of them have little or no significant role in increasing or decreasing the LDL and HDL cholesterol levels [30]. The main SFA found in *T. ferdinandiana* kernels are myristic (0.09%), palmitic (12%), stearic (7.2%), arachidic (0.76%) and behenic (0.4%) acid. The level of myristic acid in *T. ferdinandiana* kernels is only 0.09%.

Unsaturated fatty acids can exist in *cis-* or *trans*-configuration. *Cis*-configuration is found in naturally occurring unsaturated fatty acids, while *trans*-configuration is the result of processing. *Cis*-unsaturated fatty acids are known as potent inducers of adiposomes also referred to as lipid droplets and they have important roles in cell signaling, regulation of lipid metabolism and control of the synthesis and secretion of inflammatory mediators [31]. The MUFA present in *T. ferdinandiana* kernels are palmitoleate (0.2%), oleate (29.3%) and ecosenoate (0.11%). Among the MUFA, oleic acid is the most abundant one found in *T. ferdinandiana* kernels. Oleic acid has been reported to act as an anti-inflammatory and anti-apoptotic agent. The anti-inflammatory mechanism includes down-regulating cyclooxygenase-2 and inducible nitric oxide synthase through the activation of nuclear factor-kappa B [30]. Oleic acid may promote insulin resistance which is contrary to the PUFA which protects from insulin resistance [30]. Oleic acid has also been reported to attenuate blood pressure and risk of developing hypertension [32]. The potential use of *T. ferdinandiana* kernels as a dietary source of oleic acid in reducing the risk and attenuating hypertension requires further investigation. Previous reports on some of the seeds and kernels of the family Combretaceae had reported oleic acid as the most abundant unsaturated fatty acid found in this family [33].

Essential PUFA are α-linolenic (18:3, *n*-3) and linoleic acid (18:2, *n*-6), from which other important PUFA are derived. Recently, essential fatty acids (EFA) have been considered as functional food components and nutraceuticals [31]. Documented roles of EFA include cardioprotective effect (due to their considerable antiatherogenic, antithrombotic, anti-inflammatory, antiarrhythmic, hypolipidemic effects), the fluidity of biological membranes, the function of membrane enzymes and receptors, modulation of eicosanoids production, blood pressure regulation and metabolism of minerals [31]. EFA are also reported to reduce the risk of cardiovascular, cancer, osteoporosis, diabetes and some other serious diseases due to their complex effects on concentrations of lipoproteins [31]. Linoleic acid is an unsaturated omega-6 fatty acid that plays a critical role in the maintenance of the structural and functional integrity of the central nervous system (CNS) and retina [23]. A deficiency can cause skin scaling and hair loss [34]. Linoleic acid (C18:2) is the only PUFA found in *T. ferdinandiana* kernels (50.2%). Therefore, it can be suggested that *T. ferdinandiana* kernels can be used as a potential dietary source of linoleic acid which can increase the systemic pool and subsequently help nourish the CNS and retina.

The WHO recommends that total daily energy intake derived from omega-6 PUFA should be 5–8% and from omega-3 PUFA 2% for an adult male. Studies on the seeds of *T. bellirica* have reported that 40% of the seed is oil and 35% is protein and major fatty acids were linoleic (31%), palmitic (35%) and oleic (24%) acids and the authors have suggested that kernels could be used as a dietary source of linoleic acid [35]. Reported studies on various plants of *Terminalia* genus included that *T. glucausens* contains palmitic acid (34.9%), myristic acid (0.1%) and stearic acid (4.8%), seed oil of *T. superba* contains behenic acid (C22:0; 1.2%) and the oil of *T. catappa* contains stearic acid (5.8%), myristic acid (1.21%) and arachidic acid (1.3%) [33]. Variations in the fatty acid composition is very common in plants and may be due to a number of reasons including but not limited to soil composition, climate, and specific geographical locations etc.

Nutritionally, the ratio of unsaturated to saturated fatty acids in edible oils and fats is very important. High levels of saturated fatty acids are desirable to increase oil stability. However, SFA become nutritionally undesirable, because high levels of saturated fatty acids are considered to increase the concentration of LDL, affecting the ratio of LDL to HDL and promoting vascular smooth muscle proliferation [36,37]. The ratio of UFA/SFA for *T. ferdinandiana* kernels is 4, which can be considered favorable for reducing the risk of cardiovascular complications [36]. Again, the relationship between saturated and polyunsaturated FA content is an important parameter for determination of the nutritional value of oils which is expressed as P/S index. Oils and fats with a P/S index > 1 are considered to have nutritional value. Several studies indicate that a higher P/S index means a smaller deposition of lipids in the body. The P/S indexes of *T. ferdinandiana* kernels and some other common oils and fats are shown in Table 5. The P/S index of *T. ferdinandiana* kernels was 2.45, while safflower oil is 10.55 and coconut fat is 0.005. The fatty acid composition of *T. ferdinandiana* kernels is comparable to the composition of soya bean oil (Table 5).

Table 5. Comparison of the fatty acid compositions of *Terminalia ferdinandiana* kernels with commonly consumed oils and fats.

Type of Oil/Fat	Fatty Acid Composition			P/S Index	Reference
	SFA	**MUFA**	**PUFA**		
T. ferdinandiana Kernels	20.4	29.6	50.0	2.45	Current study
Coconut	90.5	8.8	0.5	0.005	
Corn	25.1	26.8	48	1.91	
Cottonseed	22.4	35.4	42	1.87	
Soybean	13.5	28.5	57.5	4.26	
Peanut	19.2	58.5	20	1.04	
Safflower	7.2	16.6	76	10.55	[36]
Linseed	9.65	22.1	68	7.05	
Palm kernel	76	22.5	1.25	0.016	
Sunflower seed	8.8	31.5	59.5	6.76	
Canola	9.6	59.5	30.7	3.2	

There are suggestions to reduce the SFA in the diet to suppress the risk of coronary heart diseases (CHD) and cardiovascular diseases (CVD). However, it is important to note that SFA reduction itself cannot suppress the risk. Mostly, reduction of SFA and TFA with their simultaneous replacement by PUFA could lead to a reduction of the risk of CHD. The SFA content of *T. ferdinandiana* kernels was 20.4%, having a very low amount of myristic acid. Based on our results, it can be concluded that the SFA of *T. ferdinandiana* kernels are unlikely to have detrimental health effects by increasing the LDL cholesterol level. Moreover, the kernels were a good source of linoleic acid suggesting *T. ferdinandiana* kernels as a valuable source of EFA that can be used in feed and food.

4. Conclusions

To the best of our knowledge, this is the first study on the nutritional composition of *T. ferdinandiana* kernels. The present study indicated that the kernels contain high levels of protein and lipid. The mineral composition of *T. ferdinandiana* kernels reveal a very good source of fundamental minerals and micronutrients. Furthermore, the kernels can be considered as a potential dietary source of linoleic and oleic acid. From a nutritional point of view, our present results suggest that *T. ferdinandiana* kernels have a high nutritional value and may contribute to a healthy diet. The utilization of kernels as a by-product from processing of *T. ferdinandiana* fruit will generate applications as a source of food supplement ingredients for essential fatty acids and can represent an alternate source of protein in the food and feed industry. Ongoing studies on the quality aspects of *T. ferdinandiana* kernels will also include bioaccessibility and bioavailability to substantiate the nutritional value and potential health effects of this unexploited by-product.

Supplementary Materials: The following are available online at http://www.mdpi.com/2304-8158/7/4/60/s1, Table S1: Trace element recoveries from reference materials.

Acknowledgments: This project is supported by RIRDC Grant (Project ID 201430161). Saleha Akter's Ph.D. is supported by Australian Government Research Training Program Scholarship.

Author Contributions: S.A. performed the experiments; collected, analyzed and interpreted the data and drafted the manuscript. Y.S., M.T.F., M.E.N. and U.T. conceived and designed the experiments, checked and approved the results and critically revised the manuscript. All authors read and approved the final version of the manuscript.

Conflicts of Interest: The authors declared no conflict of interests.

References

1. Chen, L.G.; Huang, W.T.; Lee, L.T.; Wang, C.C. Ellagitannins from *Terminalia calamansanai* induced apoptosis in HL-60 cells. *Toxicol. In Vitro* **2009**, *23*, 603–609. [CrossRef] [PubMed]

2. Hegarty, M.P.; Hegarty, E.E. Food safety of Australian Plant Bushfoods. *Rural Ind. Res. Dev. Corp.* **2001**, *28*, 33–35.

3. Singh, A.; Bajpai, V.; Kumar, S.; Kumar, B.; Srivastava, M.; Rameshkumar, K.B. Comparative profiling of phenolic compounds from different plant parts of six *Terminalia* species by liquid chromatography-tandem mass spectrometry with chemometric analysis. *Ind. Crops Prod.* **2016**, *87*, 236–246. [CrossRef]

4. Chaliha, M.; Williams, D.; Edwards, D.; Pun, S.; Smyth, H.; Sultanbawa, Y. Bioactive rich extracts from *Terminalia ferdinandiana* by enzyme-assisted extraction: A simple food safe extraction method. *J. Med. Plant Res.* **2017**, *11*, 96–106.

5. Konczak, I.; Roulle, P. Nutritional properties of commercially grown native Australian fruits: Lipophilic antioxidants and minerals. *Food Res. Int.* **2011**, *44*, 2339–2344. [CrossRef]

6. Konczak, I.; Zabaras, D.; Dunstan, M.; Aguas, P. Antioxidant capacity and hydrophilic phytochemicals in commercially grown native Australian fruits. *Food Chem.* **2010**, *123*, 1048–1054. [CrossRef]

7. Netzel, M.; Netzel, G.; Tian, Q.; Schwartz, S.; Konczak, I. Native Australian fruits—A novel source of antioxidants for food. *Innov. Food Sci. Emerg. Technol.* **2007**, *8*, 339–346. [CrossRef]

8. Shami, A.M.M.; Philip, K.; Muniandy, S. Synergy of antibacterial and antioxidant activities from crude extracts and peptides of selected plant mixture. *BMC Complement. Altern. Med.* **2013**, *13*, 360. [CrossRef] [PubMed]

9. Sultanbawa, Y.; Williams, D.; Smyth, H. *Changes in Quality and Bioactivity of Native Food during Storage*; Rural Industries Research and Development Corporation (RIRDC): Barton, Australia, 2015.

10. Sirdaarta, J.; Matthews, B.; Cock, I.E. Kakadu plum fruit extracts inhibit growth of the bacterial triggers of rheumatoid arthritis: Identification of stilbene and tannin components. *J. Funct. Foods* **2015**, *17*, 610–620. [CrossRef]

11. Tan, A.C.; Konczak, I.; Zabaras, D.; Sze, D.M.Y. Potential antioxidant, antiinflammatory, and proapoptotic anticancer activities of Kakadu plum and Illawarra plum polyphenolic fractions. *Nutr. Cancer* **2011**, *63*, 1074–1084. [CrossRef] [PubMed]

12. Williams, D.J.; Edwards, D.; Pun, S.; Chaliha, M.; Sultanbawa, Y. Profiling ellagic acid content: The importance of form and ascorbic acid levels. *Food Res. Int.* **2014**, *66*, 100–106. [CrossRef]

13. Oliveira, J.T.A.; Vasconcelos, I.M.; Bezerra, L.; Silveira, S.B.; Monteiro, A.C.O.; Moreira, R.A. Composition and nutritional properties of seeds from *Pachira aquatica* Aubl, *Sterculia striata* St Hil et Naud and *Terminalia catappa* Linn. *Food Chem.* **2000**, *70*, 185–191. [CrossRef]

14. Akpakpan, A.E.; Akpabio, U.D. Evaluation of Proximate composition, mineral element and anti-nutrient in Almond (*Terminalia catappa*) seeds. *Res. J. Appl. Sci.* **2012**, *7*, 489–493.

15. Ladele, B.; Kpoviessi, S.; Ahissou, H.; Gbenou, J.; Kpoviessi, B.K.; Mignolet, E.; Herent, M.F.; Bero, J.; Larondelle, Y.; Leclercq, J.Q.; et al. Chemical composition and nutritional properties of *Terminalia catappa* L. oil and kernels from Benin. *Com. Ren. Chim.* **2016**, *19*, 876–883. [CrossRef]

16. Cheng, J.; Zhou, S.; Wu, D.; Chen, J.; Liu, D.; Ye, X. Bayberry (*Myrica rubra* Sieb. et Zucc.) kernel: A new protein source. *Food Chem.* **2009**, *112*, 469–473. [CrossRef]

17. Chivandi, E.; Mukonowenzou, N.; Berliner, D. The coastal red-milkwood (*Mimusops caffra*) seed: Proximate, mineral, amino acid and fatty acid composition. *S. Afr. J. Bot.* **2016**, *102*, 137–141. [CrossRef]

18. Lv, Z.C.; Chen, K.; Zeng, Y.W.; Peng, Y.H. Nutritional composition of *Canarium pimela* L. kernels. *Food Chem.* **2011**, *125*, 692–695. [CrossRef]

19. He, Z.; Xia, W. Nutritional composition of the kernels from *Canarium album* L. *Food Chem.* **2007**, *102*, 808–811. [CrossRef]

20. Poudyal, H.; Kumar, S.A.; Iyer, A.; Waanders, J.; Ward, L.C.; Brown, L. Responses to oleic, linoleic and α-linolenic acids in high-carbohydrate, high-fat diet-induced metabolic syndrome in rats. *J. Nutr. Biochem.* **2013**, *24*, 1381–1392. [CrossRef] [PubMed]

21. Tinggi, U.; Schoendorfer, N.; Scheelings, P.; Yang, X.; Jurd, S.; Robinson, A.; Smith, K.; Piispanen, J. Arsenic in rice and diets of children. *Food Addit. Contam.* **2015**, *8*, 149–156. [CrossRef] [PubMed]

22. Carter, J.F.; Tinggi, U.; Yang, X.; Fry, B. Stable isotope and trace metal compositions of Australian prawns as a guide to authenticity and wholesomeness. *Food Chem.* **2015**, *170*, 241–248. [CrossRef] [PubMed]

23. Chivandi, E.; Davidson, B.C.; Erlwanger, K.H. Proximate, mineral, fibre, phytate-phosphate, vitamin E, amino acid and fatty acid composition of *Terminalia sericea*. *S. Afr. J. Bot.* **2013**, *88*, 96–100. [CrossRef]

24. Phillips, S.M. Dietary protein requirements and adaptive advantages in atheletes. *Br. J. Nutr.* **2012**, *108*, S158–S167. [CrossRef] [PubMed]

25. Mehmood, S.; Orhan, I.; Ahsan, Z.; Aslan, S.; Gulfraz, M. Fatty acid composition of seed oil of different *Sorghum bicolor* varieties. *Food Chem.* **2008**, *109*, 855–859. [CrossRef] [PubMed]

26. Cheng, J.; Ye, X.; Chen, J.; Liu, D.; Zhou, S. Nutritional composition of underutilized bayberry (*Myrica rubra* Sieb. et Zucc.) kernels. *Food Chem.* **2008**, *107*, 1674–1680. [CrossRef]

27. Chung, S.W.C.; Kwong, K.P.; Yau, J.C.; Wong, W.W. Dietary exposure to antimony, lead and mercury of secondary school students in Hong Kong. *Food Addit. Contam.* **2008**, *25*, 831–840. [CrossRef] [PubMed]

28. Patil, V.; Gislerød, H.R. The importance of omega-3 fatty acids in diet. *Curr. Sci.* **2006**, *90*, 908–909.

29. Iso, H.; Sato, S.; Umemura, U.; Kudo, M.; Koike, K.; Kitamura, A.; Imano, H.; Okamura, T.; Naito, Y.; Shimamoto, T. Linoleic acid, other fatty acids, and the risk of stroke. *Stroke* **2002**, *33*, 2086–2093. [CrossRef] [PubMed]

30. Orsavova, J.; Misurcova, L.; Ambrozova, J.; Vicha, R.; Mlcek, J. Fatty Acids Composition of Vegetable Oils and Its Contribution to Dietary Energy Intake and Dependence of Cardiovascular Mortality on Dietary Intake of Fatty Acids. *Int. J. Mol. Sci.* **2015**, *16*, 12871–12890. [CrossRef] [PubMed]

31. Nantapo, C.W.T.; Muchenje, V.; Nkukwana, T.T.; Hugo, A.; Descalzo, A.; Grigioni, G.; Hoffman, L.C. Socio-economic dynamics and innovative technologies affecting health-related lipid content in diets: Implications on global food and nutrition security. *Food Res. Int.* **2015**, *76*, 896–905. [CrossRef]

32. Alonso, A.; Martinez-González, M.A. Olive oil consumptionand reduced incidence of hypertension: The SUN study. *Lipids* **2004**, *39*, 1233–1238. [CrossRef] [PubMed]

33. Balogun, A.M.; Fetuga, B.L. Fatty Acid Composition of Seed Oils of Some Members of the Meliaceae and Combretaceae Families. *J. Am. Oil Chem. Soc.* **1985**, *62*, 529–531. [CrossRef]

34. Khodadoust, S.; Mohammadzadeh, A.; Mohammadi, J.; Irajie, C.; Ramezani, M. Identification and determination of the fatty acid composition of Quercus brantii growing in southwestern Iran by GC–MS. *Nat. Prod. Res.* **2014**, *28*, 573–576. [CrossRef] [PubMed]

35. Rukmini, C.; Rao, P.U. Chemical and Nutritional Studies on *Terminalia bellirica* Roxb. Kernel and Its Oil. *J. Am. Oil Chem. Soc.* **1986**, *63*, 360–363. [CrossRef]

36. Kostik, V.; Memeti, S.; Bauer, B. Fatty acid composition of edible oils and fats. *J. Hyg. Eng. Des.* **2013**, *4*, 112–116.

37. He, Z.; Zhu, H.; Li, W.; Zeng, M.; Wu, S.; Chen, S.; Qin, F.; Chen, J. Chemical components of cold pressed kernel oils from different *Torreya grandis* cultivars. *Food Chem.* **2016**, *209*, 196–202. [CrossRef] [PubMed]

Review

Phytochemical Properties and Nutrigenomic Implications of Yacon as a Potential Source of Prebiotic: Current Evidence and Future Directions

Yang Cao [1,*,†], Zheng Feei Ma [2,3,*,†], Hongxia Zhang [4,†], Yifan Jin [2], Yihe Zhang [5] and Frank Hayford [6]

[1] Department of Health Promotion, Pudong Maternal and Child Health Care Institution, Shanghai 201399, China
[2] Department of Public Health, Xi'an Jiaotong-Liverpool University, Suzhou 215123, China; Yifan.Jin14@student.xjtlu.edu.cn
[3] School of Medical Sciences, Universiti Sains Malaysia, Kota Bharu 15200, Kelantan, Malaysia
[4] Department of Food Science, University of Otago, Dunedin 9016, New Zealand; zhanghongxia326@hotmail.com
[5] Division of Medicine, School of Life and Medical Sciences, University College London, London WC1E6BT, UK; yihe.zhang.16@ucl.ac.uk
[6] Department of Nutrition and Dietetics, School of Biomedical and Allied Health Sciences, College of Health Sciences, University of Ghana, Accra P.O. Box LG 25, Ghana; feahayford220580@gmail.com
* Correspondence: evacaoyang@163.com (Y.C.); Zhengfei.Ma@xjtlu.edu.cn (Z.F.M.); Tel.: +86-21-580-26156 (Y.C.); +86-512-8188-4938 (Z.F.M.)
† These authors contributed equally to this work and shared the co-first authorship.

Received: 5 March 2018; Accepted: 9 April 2018; Published: 12 April 2018

Abstract: The human gut is densely populated with diverse microbial communities that are essential to health. Prebiotics and fiber have been shown to possess the ability to modulate the gut microbiota. One of the plants being considered as a potential source of prebiotic is yacon. Yacon is an underutilized plant consumed as a traditional root-based fruit in South America. Yacon mainly contains fructooligosaccharides (FOS) and inulin. Therefore, it has bifidogenic benefits for gut health, because FOS are not easily broken down by digestive enzymes. Bioactive chemical compounds and extracts isolated from yacon have been studied for their various nutrigenomic properties, including as a prebiotic for intestinal health and their antimicrobial and antioxidant effects. This article reviewed scientific studies regarding the bioactive chemical compounds and nutrigenomic properties of extracts and isolated compounds from yacon. These findings may help in further research to investigate yacon-based nutritional products. Yacon can be considered a potential prebiotic source and a novel functional food. However, more detailed epidemiological, animal, and human clinical studies, particularly mechanism-based and phytopharmacological studies, are lacking for the development of evidence-based functional food products.

Keywords: yacon; *Smallanthus sonchifolius*; prebiotic; fructooligosaccharides; underutilized

---

1. Introduction

A focus on the role of gut microbiota to improve health and prevent disease has attracted intense interest in identifying dietary strategies to modulate the gut microbiota. One such dietary strategy includes the intake of prebiotics and dietary fiber, because they can be metabolized by the gut microbiota. One potential candidate for prebiotics is yacon, which has an abundance of free sugar and fructans with low polymerization (i.e., fructooligosaccharides (FOS)).

Yacon (*Smallanthus sonchifolius*), an underutilized crop, belongs to the family Asteraceae [1]. Originating from the Andean region of South America, yacon is a little known perennial herb that generally takes 6 to 12 months to reach maturity. The aerial stems of yacon can reach about 2.5 m in height [1]. The roots of yacon have a similar appearance to sweet potato, and their weight is about 500 g each. Each yacon plant has about 20 units and can yield >10 kg of roots [2]. The roots of yacon are about 10 cm thick and 20 cm long, in various sizes and shapes. Yacon tuber roots are usually eaten as fruits [3]. The edible roots of yacon are juicy and sweet like an apple and can be consumed either raw or cooked. Its leaves can also be used to brew a medicinal tea [4].

Since the pre-Incan period, yacon has been cultivated and consumed. Its low nutritive value might be one of the reasons why it is being neglected, especially by older Andean agronomists [1]. The scientific community paid little attention to yacon until the 1980s, except in Peru and Japan [5]. Yacon products can range from pickles to dried flakes. Although yacon has a sweet taste and is juicy, it is considered a food with low energy value because of the low-molecular-weight carbohydrate FOS [1]. The roots of yacon mainly contain fructans, and its leaves have been reported to possess putative medicinal compounds. Yacon can provide fiber and low calories for consumers who have an inactive lifestyle with excess intake of fats and carbohydrates. Also, the roots of yacon lack starch, which might be beneficial for the diets of diabetes patients [6]. Therefore, yacon actually has great potential to be a useful plant species.

Its growth and cultivation have spread widely to several countries, such as the Czech Republic, China, Brazil, Japan, Italy, and New Zealand, in recent years due to its presumed physiological benefits and high adaptation to different cultivation environments [3]. The global expansion of yacon cultivation and marketing was further motivated after studies reported on the health benefits of consuming yacon, such as the antioxidant activity associated with its phenolic compounds and the reduction of blood glucose level attributed to its carbohydrate profile [3,6].

Several bioactive compounds were found in both the roots and leaves of yacon, including polyphenol compounds, fructans, and phytoalexins, which show antioxidant, prebiotic, and antimicrobial properties [7,8]. Carbohydrates are stored in yacon in the form of β-(2→1) FOS that can help to prevent constipation and reduce the concentrations of blood glucose and lipids [9]. These functional properties could help people maintain health and reduce the risk of chronic diseases [10,11]. This review will analyze the accumulated evidence on the phytochemical compounds and nutrigenomic properties of yacon, paying special attention to its role as a prebiotic.

Search Strategy

An electronic literature search was conducted using Cochrane Library, Medline (OvidSP), Google Scholar, and PubMed through January 2018. Additional articles were identified from references located in the retrieved articles. Our search strategy included combinations of the following using Boolean markers: *Smallanthus sonchifolius*, yacon, health, prebiotic, phytochemicals, and fructans. The search was restricted to experimental, epidemiological, and clinical studies published in English that address the phytochemical constituents and nutrigenomic properties of yacon. Our work will also add to the current understanding of some of the bioactive compounds and nutrigenomic properties of yacon that were covered adequately in the previous reviews [2].

2. Phytochemical Compounds

2.1. Roots/Tubers

Table 1 shows the chemical composition of fresh yacon root. There are three main substances in fresh yacon root: water (>70%), carbohydrates (the major proportion of the dry matter), and protein [12]. Carbohydrates in yacon root contain glucose, fructose, sucrose, and FOS. Among them, FOS are considered the predominant saccharides [4]. FOS are natural food components found in many plants. However, FOS concentrations in the roots of yacon are highest compared to other plants [13].

The chemical structure of FOS has 2 to 10 fructose molecules connected with a β-(1,2) glucosidic bond and 1 glucose molecule linked with α-(1,2) bond [14]. FOS are stable under conditions of pH > 3 and temperature up to 140 °C [15]. The major FOS in yacon include nystose, 1-kestose, and 1-fructofuranosyl nystose [16]. The carbohydrate content in yacon can be influenced by the location of cultivation, season of growing, and time and temperature of postharvest storage. With increased time after harvesting, fructans in yacon rapidly depolymerize to mono- and disaccharides by fructan hydrolase. Under a low temperature of postharvest storage (~10 °C), this conversion speed is slower [1]. FOS is nondigestible in the upper gastrointestinal tract before going through fermentation in the large intestine. The small intestine does not have enzymes to hydrolyze the glucosidic bonds in FOS. Studies have shown that FOS can be fermented by most *Bifidobacterium* strains and some *Lactobacillus* strains, healthy beneficial bacteria that naturally exist in the colon [17,18]. A study by Pedreschi et al. [17] indicated that both *Lactobacillus* and *Bifidobacterium* strains utilized GF2 in root extracts of yacon, while *Bifidobacterium* utilized molecules with longer FOS chains. Consumption of FOS could produce short-chain fatty acids and lead to an increase in *Bifidobacteria* [19–21]. A clinical study by Guigoz [22] showed that FOS consumption could modulate intestinal microbiota and have a beneficial effect in improving health outcomes. Taken together, FOS are considered as a prebiotic that meets the criteria defined by the Food and Agriculture Organization (FAO) on prebiotics [23]. Furthermore, FOS have been reported to increase bone density and absorption of magnesium and calcium [24–26]. In addition, free fructose is naturally present in vegetables and fruits, including yacon, so its intake is an unavoidable consequence of eating a healthy diet. Only when fructose intake is excessive does it have deleterious metabolic effects in humans [27].

Table 1. Chemical composition of 1 kg fresh yacon root [13].

Variables	Mean
Dry matter (g)	115
Total carbohydrates (g)	106
Fructans (g)	62
Total free sugars (g)	26
Free glucose (g)	3.5
Free sucrose (g)	14
Free fructose (g)	8.5
Protein (g)	3.7
Fiber (g)	3.6
Fat (mg)	244
Calcium (mg)	87
Potassium (mg)	2282
Phosphorus (mg)	240

In a study by Goto et al. [12], researchers purified and confirmed the presence of oligosaccharides in the roots of yacon as β-(2→1) with terminal sucrose, which are inulin-type oligofructans, using enzymatic, ^{13}C-nuclear magnetic resonance (NMR), and methylation methods. Yan et al. [6] showed the presence of chlorogenic acid and tryptophan in the roots of yacon using NMR and mass spectrometry. Another study, by Takenaka et al. [28], identified five caffeic acid derivatives in the roots of yacon using spectroscopic methods. These compounds were 2,5-dicaffeoylaltraric acid, 3,5-dicaffeoylquinic acid, chlorogenic acid (3-caffeoylquinic acid), 2,4- or 3,5-dicaffeoylaltraric acid, and 2,3,5- or 2,4,5-tricaffeoylaltraric acid [28].

In addition, flavonoids were found only in acid-hydrolyzed yacon tubers and were found to have a relationship with lipid peroxidation, acetylcholinesterase, and butyrylcholinesterase inhibition [5,29]. A study by Simonovska et al. [5] also showed the presence of the flavonoid quercetin, caffeic acid, and ferulic acid using thin-layer chromatography in the acid hydrolysis of yacon tubers. Also, yacon root contains small amounts of vitamins and minerals. Among them, vitamin C and potassium are the most abundant nutrients [15]. Tryptophan, known as a precursor of serotonin and melatonin, is the most abundant amino acid in yacon root [6]. Tryptophan has antioxidant properties. It has been observed that tryptophan is more likely to eliminate free radicals from the oxidative damage of low-density lipoprotein compared with melatonin [30]. However, tryptophan has less antioxidant activity than chlorogenic acid by 1,1-diphenyl-2-picrylhydrazyl (DPPH) assay [6].

2.2. Leaves and Flowers

Lin et al. [7] extracted and reported antibacterial compounds including 8β-tigloyloxymelampolid-14-oic acid methyl ester, melampolide-type sesquiterpene lactones, and 8β-methacryloyloxymelampolid-14-oic acid methyl ester, uvedalin, sonchifolin, fluctuanin, melampolides, and enhydrin from yacon leaves. Simonovska et al. [5] also reported the presence of ferulic acid in yacon leaves using thin-layer chromatography.

Yacon is also rich in polyphenols. A study by Hondo et al. [31] showed that yacon juice had 850 ppm of phenolic compounds. A higher polyphenol concentration is usually found in leaves and stems. Chlorogenic, caffeic, ferulic, and protocatechuic acids in tuber and leaf extracts of yacon have also been detected by thin-layer chromatographic screening [5]. Among them, chlorogenic and caffeic acid derivatives are the main polyphenols in yacon, the former at a higher concentration [6]. One investigation of phenolic compounds of yacon found that it contained five caffeic acid derivatives [28]. Once yacon tissue is exposed to the air, it will darken rapidly. The browning reaction is due to a condensation reaction of phenolic compounds with the enzymatic polymerization of polyphenols and amino acids [4]. Polyphenols in yacon leaves provide an acrid and astringent flavor and characteristic odor. Polyphenols are highly related to superoxide radical, DPPH radical, and nitric oxide scavenging activities, which indicates that these compounds have antioxidant properties and may play an important role in lowering the risk of cancer, cardiovascular disease (CVD), atherosclerosis, and diabetes [10,29,32,33].

Sonchifolin, polymatin B, uvedalin, two melampolide-type sesquiterpene lactones, and enhydrin were isolated from yacon leaf extract as an antifungal substance; among them, sonchifolin showed high antimicrobial activity against *Pyricularia oryzae* [7,34]. Moreover, it was reported that there is a high proportion of *ent*-kaurenic acid and kaurene derivatives in yacon leaves and they are involved in the protective mechanism of the glandular trichome exudates [35]. Table 2 shows an overview of major phytochemical compounds in yacon.

Table 2. An overview of major phytochemical compounds of yacon.

Parts of Yacon	Compounds/Nutrients Identified	Test Methods	References
Roots/tubers	Fructooligosacharides (1-kestose, nystose, and 1-fructofuranosyl nystose)	Fermentation by *Bifidobacterium and Lactobacillus*	Hermann et al. (1997) [13]; Niness et al. (1999) [14]; Roberfroid et al. (2010) [16]; Delgado et al. (2012) [4]; Paula et al. (2014) [15]
	Tryptophan	2,2′-azino-bis(3-ethylbenzothiazoline-6-sulphonic acid) (ABTS)	Sousa et al. (2015) [36]
	Chlorogenic acid	ABTS	Sousa et al. (2015) [36]
	Caffeic acid	ABTS	Sousa et al. (2015) [36]
	Ferulic acid	1,1-diphenyl-2-picrylhydrazyl (DPPH)	Simonovska et al. (2003) [5]
Leaves	Chlorogenic acid	Decoction DPPH and xanthine/xanthine oxidase (XOD) superoxide radical scavenging assays Ohmic-assisted decoction	Yan et al. (1999) [6]; Genta et al. (2009) [37]; Valentová et al. (2003) [9]; Simonovska et al. (2003) [5]; Khajehei et al. (2017) [38]

Table 2. *Cont.*

Parts of Yacon	Compounds/Nutrients Identified	Test Methods	References
	Caffeic acid	Decoction DPPH and xanthine/XOD superoxide radical scavenging assays Ohmic-assisted decoction	Genta et al. (2009) [37]; Russo et al. (2015) [29]; Valentová et al. (2003) [10]; Khajehei et al. (2017) [38]
	Ferulic acid	DPPH and xanthine/XOD superoxide radical scavenging assays Ohmic-assisted decoction	Valentová et al. (2003) [10]; Khajehei et al. (2017) [38]
	Myricetin	Ohmic-assisted decoction	Khajehei et al. (2017) [26]
	Rutin	Decoction Ohmic-assisted decoction	De Andrade et al. (2014) [3]; Khajehei et al. (2017) [38]
	p-Coumaric acid	Ohmic-assisted decoction	Khajehei et al. (2017) [38]
	Gallic acid	Decoction	De Andrade et al. (2014) [3]
	Tryptophan	DPPH assay	Yan et al. (1999) [6]
	Enhydrin	Decoction	Genta et al. (2009) [37]
Flower	Myricetin	Decoction	De Andrade et al. (2014) [3]
	Gallic acid	Decoction	De Andrade et al. (2014) [3]

3. Nutrigenomic Properties of Yacon

Since yacon is an underutilized plant, limited studies have been conducted to determine its nutrigenomic properties [1]. Therefore, there is no conclusive information on the relationship between yacon consumption and its nutrigenomic value. Although the literature shows an association with the nutrigenomic properties [1,39], a causal relationship between yacon and observed health outcomes has not been firmly established. Similar to nutrigenomic properties reported in other plants [40–42], the findings of such studies on yacon should be interpreted with caution. Table 3 shows the major nutrigenomic properties of yacon. Yacon has been investigated for its various nutrigenomic properties using epidemiological, animal, and human clinical studies. In addition, the possible mechanisms underlying some of them have been determined and are discussed in the following section.

Table 3. An overview of major nutrigenomic properties of yacon.

Pharmacological Effects	Models Used	Parts/Forms of Yacon Used	References
Hypoglycemia effect	Streptozotocin-induced diabetic rats	Leaf extract	Aybar et al. (2001) [43]; Valentová and Ulrichová (2003) [9]
	Diabetic rats	Aqueous leaf extract	Simonovska et al. (2003) [5]; Barcellona et al. (2012) [39]
	Streptozotocin-induced diabetic and nondiabetic rats	Leaf extract	Baroni et al. (2008) [44]
	Streptozotocin-induced diabetic rats	Dried root extract	Satoh et al. (2013) [45]; Oliveira et al. (2016) [46]
	Normoglycemic, transiently hyperglycemic, and diabetic rats	Leaf extract	Genta et al. (2009) [37]
	Decoction and enhydrin-fed Wistar rats	Leaf extract	Barcellona et al. (2012) [39]
	Humans	Freeze-dried powder	Scheid et al. (2014) [11]
	Humans	Syrup	Genta et al. (2009) [37]
Hypolipidemic effect	Normal and streptozotocin-induced diabetic rats	Dried root flour	Genta et al. (2005) [47]; Habib et al. (2011) [48]
	Hypercholesterolemic male Wistar rats	Root extract	Oliveira et al. (2016) [46]; Oliveira et al. (2013) [49]
	Mildly dyslipidemic premenopausal women		Genta et al. (2009) [37]

Table 3. *Cont.*

Pharmacological Effects	Models Used	Parts/Forms of Yacon Used	References
Anti-inflammatory effects	Hypercholesterolemic rats	Root extract	Oliveira et al. (2016) [46]
	Adult male BALB/c mice	Leaf extract	Inoue et al. (1995) [34]; Lin et al. (2003) [7]; Schorr et al. (2007) [50]; Oliviera et al. (2013) [49]
Antimicrobial effects	*P. oryzae*	Leaf extract	Inoue et al. (1994) [34]
	B. subtilis	Leaf extract	Lin et al. (2003) [7]
	Gram-positive organisms (*Staphylococcus aureus, Staphyilococcus epidermidis, and Bacillus subtilis*)	Leaf extract	Padla et al. (2012) [51]

3.1. Beneficial Effects on Intestinal Health

Colorectal cancer (CRC) is the third leading cause of cancer deaths globally [52]. Many risk factors are linked with the occurrence of this disease. Sporadic lifestyle and dietary habits are the main risk factors for most CRC cases [53,54]. Constipation is positively related with an increased risk of colon cancer [55]. In one study, intestinal transit time was significantly decreased, with a slight increase of stool frequency and a tendency for softer stools, after the consumption of 20 g yacon syrup for 2 weeks among healthy participants (n = 16) [8]. An improvement in bowel movements was found in a study of constipated elderly patients (n = 5) [56] and a study of women with a history of constipation (n = 55) [37] that used the same dose of 0.14 g FOS/kg of body weight. These findings [37,56] suggest that foods rich in FOS such as yacon may improve constipation.

A study on 1,2-dimethylhydrazine (DMH)-induced models of colon carcinogenesis in rats showed that yacon and a symbiotic formulation (yacon plus *Lactobacillus acidophilus*) were associated with a reduction of cell proliferation and tumor multiplicity [57]. Similar findings were found in a more recent study, which also showed that aqueous extracts of yacon significantly decreased DNA damage in leukocytes of DMH-induced rats [15].

The mechanisms of improving intestine health might be due to high amounts of FOS in yacon. FOS can stimulate the growth of *Bifidobacteria* to inhibit the growth of pathogenic bacteria. In addition, increasing short-chain fatty acids (SCFAs) produced by FOS fermentation can activate the immune response, lower the pH in the colon, and promote the excretion of amine and ammonia [58]. With reference to carcinogenesis, SCFAs reduce cellular proliferation and cause apoptosis. Studies in rats showed that butyrate slowed down the progress of preneoplastic aberrant crypt foci lesions and postponed the development of tumors [57,59,60].

In an intervention study investigating the effect of yacon flour on nutritional status and immune response biomarkers, Vaz-Tostes et al. [61] reported that preschool children aged 2 to 5 years had an improved intestinal immune response, as shown by an increase in the concentration of serum interleukin (IL)-4 and secretory IgA (sIGA) after the intervention. However, no improvement was seen in the biomarkers of zinc and iron in children [61].

3.2. Hypoglycemia Effect

Several studies have shown that yacon has a beneficial effect on reducing blood sugar [43–45]. For example, in an animal study, after 30 days, the glucose levels of streptozotocin (STZ)-induced diabetic rats significantly decreased when the rats were treated with 2% yacon tea administered ad libitum [43]. Meanwhile, the plasma insulin level was improved in the treated group as well [43]. These findings were supported by another research study that observed a significant reduction of glycemia in STZ-induced nondiabetic and diabetic rats when they were fed leaf extracts of yacon obtained by hydro-ethanolic extraction [44]. However, when using extracts of yacon obtained by other extract solutions, no hypoglycemic effects were reported, implying that the method of obtaining the extracts was noteworthy. Similar results were found when the experimental material was replaced by dried root extracts [45].

In a 120-day, double-blind, placebo-controlled human intervention study, the consumption of yacon syrup significantly decreased the homeostasis model assessment for insulin resistance and fasting serum insulin in women who were dyslipidemic, premenopausal, and obese [37]. Among elderly individuals, a 9-week intake of freeze-dried powder of yacon was associated with lower serum glucose levels [11]. No significant changes were reported for insulin-stimulated glucose metabolism and fasting plasma glucose of healthy participants when consuming 20 g FOS per day [62].

3.3. Hypolipidemic Effect

A study by Genta et al. [47] reported that oral intake of dried yacon root flour for 4 months significantly reduced serum triacylglycerol (TG) levels in normal rats. Their findings were corroborated by a study that observed significant decreases in serum TG levels and very-low-density lipoprotein (VLDL) in STZ-induced diabetic rats treated with yacon flour for 90 days [48]. Interestingly, the low dose of FOS (340 mg/kg) showed more hypolipidemic effect than the high dose (6800 mg/kg). Moreover, Oliveira et al. [46] found that the concentrations of high-density lipoprotein (HDL) cholesterol and total cholesterol were significantly improved in male Wistar rats fed yacon extracts for 14 days. Another study [63] reported that yacon-supplemented diabetic rats had lower malondialdehyde levels in both liver and kidney. Also, yacon-supplemented diabetic rats had lower hepatic dismutase and catalase activity than the controls [63].

Overall, the results in animal studies [46–48] seem to be more convincing, while human studies are more controversial [37]. As mentioned previously, a significant reduction of low-density lipoprotein (LDL) was seen in the study of mildly dyslipidemic premenopausal women [37]. These findings were in accordance with results indicating that the hypolipidemic effect of inulin-type fructans is mostly observed in dyslipidemia patients [64]. However, in a double-blind, placebo-controlled study, no reduction in serum lipid levels was reported in an elderly population (n = 72) supplemented with freeze-dried yacon powder for 9 weeks [11].

3.4. Anti-Inflammatory Effect

Studies have demonstrated that yacon also possesses anti-inflammatory action [7,34,49,50,65,66]. For example, yacon leaf extract might be used as a promising therapeutic agent, especially in topical applications. It is suggested that the anti-inflammatory activity is associated with sesquiterpene lactones (STLs) [7,49], which are found in higher concentrations in the leaves of yacon [34,49,67]. Uvedalin, enhydrin, sonchifolin, and polimatin B are the main STLs detected in yacon leaves [50]. Enhydrin and uvedalin, for instance, have anti-inflammatory properties, as shown by their inhibition of transcription factor NF-κB [50]. A study by Oliveira et al. [49] reported that yacon extract had topical anti-inflammatory and anti-edematous activity, because it reduced edema and neutrophil migration to inflammatory sites when administered to adult male BALB/c mice used as test subjects. This activity may be an important part of the anti-inflammatory action of the extract, exerting some effects on inflammatory mediators, thus demonstrating that yacon leaf extract possesses topical anti-edematous activity in vivo and can be developed as a topical anti-inflammatory agent [49].

3.5. Antioxidant Activity

Oxidative stress is suspected to be involved in many chronic diseases, including neurodegenerative diseases, cardiovascular diseases, and certain age-related cancers [36]. Several studies have reported on the presence of phenolic compounds, including caffeic acid, chlorogenic acid, and ferulic acid, which are known to be natural dietary antioxidants, in leaf and tuber extracts of yacon [5,9,28,33]. A study by Oliveira et al. [68] reported that male Wistar rats fed yacon extract had a significant reduction in the serum levels of cardiac markers and an increase in antioxidant defense. In a study investigating the antioxidant properties of sterilized yacon tuber flour, the antioxidant activity of yacon extract was tested by biological assays to determine the effects of protection on directly exposed and phagic DNA [36]. The study results showed that yacon extract had antioxidant activity in protecting DNA from oxidative degradation in both situations, which was contributed by its phenolic compound composition [36].

In a study determining the in vivo antioxidant action of yacon extracts [69], when rat hepatocyte primary cultures were preincubated with yacon leaf extracts before oxidative damage induced by allyl alcohol and tert-butyl hydroperoxide, the toxic effect was less pronounced and the hepatocytes retained high viability. This study indicated that all yacon extracts tested had significant effects on radical scavenging and strong protective effects against oxidative damage to rat hepatocytes [69]. In addition, the study suggested that yacon extracts had an effect on reducing hepatic glucose production, which might contribute to the prevention and treatment of diabetes [69].

3.6. Antimicrobial Properties

The cultivation of yacon usually requires almost no pesticides, suggesting that it naturally possesses antimicrobial substances [7]. For example, Inoue et al. [34] first isolated a new antifungal melampolide and three known melampolides from yacon extracts as fungicidal compounds against *P. oryzae*.

Lin et al. [7] also found six melampolide-type sesquiterpene lactones, which were categorized as antibacterial compounds based on their inhibition of *B. subtilis*. A study by Padla et al. [51] investigated one of the antimicrobial compounds isolated from yacon extracts, ent-kaurenoic, which was shown to be active against gram-positive organism (*S. aureus*, *S. epidermidis*, and *B. subtilis*) at the lowest concentration (1000 µg/mL). The authors [51] also suggested that yacon extract has potential protective effects on bacterial skin infections due to its anti-staphylococcal properties. However, more evidence is needed to establish the antimicrobial effects of yacon, because little is known about the antimicrobial effects of yacon on the gut microbiota.

3.7. Beneficial Effects on Minerals Balance

Since yacon contains high concentration of fructans, especially FOS, which are regarded as prebiotic ingredients, FOS could be selectively fermented by the gut microbiota in the large intestine. Consequently, this will result in increased levels of short chain fatty acids (SCFAs). SCFAs have an effect on lowering the luminal pH, thereby increasing the solubility of minerals and absorption in the large intestine [70,71].

Lobo et al. [72] conducted a study in growing rats to evaluate the effects of yacon flour consumption on calcium and magnesium balance and bone health. Their results showed that taking yacon flour as a dietary supplement significantly improved intestinal absorption and calcium and magnesium balance, resulting in higher bone mineral retention and stronger bone structural properties than the control group. Nevertheless, increasing the concentration of FOS to 5% or 7.5% in yacon flour fed to rats was related to a significant increase in calcium absorption and calcium and magnesium balance. Although all bone parameters showed an increase in the yacon-supplemented group compared with the control group, only peak loads and stiffness were observed to significantly differ between the groups [72], which may be due to the increasing number and depth of bifurcated crypts. Another study, by Rodrigues et al. [73], reported that rats fed with yacon flour plus *Bifidobacterium longum* (*B. longum*) had significantly higher tibia mineral content (calcium, magnesium, and phosphorus) and fracture strength.

As for human studies, the calcium concentration in blood was observed to be increased in the intervention group receiving yacon syrup as a secondary outcome in the study of Genta et al. [37]. However, there are inconsistencies in the effect of FOS on calcium absorption. Some studies have reported a positive effect of FOS on stimulating calcium absorption in adolescents, young men, and pre- and postmenopausal women [37,74], while one study showed no effect of consuming 15 g of FOS/day for 21 days in a group of healthy young men (*n* = 12) [75]. A possible reason for such discrepancies might be the small doses of FOS and inappropriate methods used to determine calcium absorption [75].

3.8. Adverse Effects

A study reported that subjects had side effects such as diarrhea, flatulence, nausea, and abdominal distension when consuming yacon at 0.29 g FOS/kg body weight/day [37]. These symptoms disappeared when the dose was reduced to 0.14 g FOS/kg body weight/day. In an animal study, no differences were found between low and high concentrations in the yacon-supplemented group with regard to adverse co5nsequences, except that cecal hypertrophy was observed in a few rats in the high-concentration yacon group [47]. There were two cases of severe adverse effects after consuming yacon. One was the case of a 55-year-old woman who suffered from anaphylaxis after ingesting yacon root [76], and the other was an animal study [77] that reported the development of renal lesions in rats with long-term consumption of yacon leaves.

Indicators of liver function in rats showed no significant difference after 28 days of a yacon diet, which indicates an absence of liver toxicity due to supplementation with yacon [73]. These findings were consistent with a study of 4 months of yacon flour supplementation (0.6% and 13% FOS) in rats [47].

4. Conclusions

Yacon has multiple nutrigenomic implications with regard to health outcomes. In addition, yacon is a useful resource for alternative and complementary prebiotics, for example, for intestinal health and for their antimicrobial and antioxidant effects. Although yacon has a long history as a root-based fruit in South America, future studies are needed to better elucidate its mechanisms and nutrigenomic properties regarding health outcomes. This is because there are limited data, especially on the safety evaluation of yacon. In addition, having a better understanding of the effects and mechanisms involved would allow yacon to be developed as a novel functional food as well.

Acknowledgments: Zheng Feei Ma would like to thank Siew Poh Tan, Peng Keong Ma, Siew Huah Tan, Feng Yuan Lau, and Zheng Xiong Ma for their active encouragement and support of this work. The authors received no specific funding for this work.

Author Contributions: The project idea was developed by Z.F.M. Y.C. and Z.F.M. wrote the first draft of the manuscript. Y.C., Z.F.M., H.Z., Y.J., Y.Z., and F.H. conducted the literature review and revised the manuscript.

Conflicts of Interest: The authors declare no conflict of interest.

References

1. Grau, A.; Rea, J. Yacon. Smallanthus Sonchifolius (Poepp. & Endl.) H. Robinson. In *Andean Roots and Tuberous Roots: Ahipa, Arracacha, Maca and Yacon. Promoting the Conservation and Use of Underulitized Crops*; Hermann, M., Heller, J., Eds.; IPK: Rome, Italy, 1997; pp. 199–256.
2. Ojansivu, I.; Ferreira, C.L.; Salminen, S. Yacon, a new source of prebiotic oligosaccharides with a history of safe use. *Trends Food Sci. Technol.* **2011**, *22*, 40–46. [CrossRef]
3. De Andrade, E.F.; Leone, R.D.S.; Ellendersen, L.N.; Masson, M.L. Phenolic profile and antioxidant activity of extracts of leaves and flowers of yacon (*Smallanthus sonchifolius*). *Ind. Crops Prod.* **2014**, *62*, 499–506. [CrossRef]
4. Delgado, G.T.; Tamashiro, W.M.; Marostica Junior, M.R.; Pastore, G.M. Yacon (*Smallanthus sonchifolius*): A functional food. *Plant Foods Hum. Nutr.* **2013**, *68*, 222–228. [CrossRef] [PubMed]
5. Simonovska, B.; Vovk, I.; Andrensek, S.; Valentova, K.; Ulrichova, J. Investigation of phenolic acids in yacon (*Smallanthus sonchifolius*) leaves and tubers. *J. Chromatogr. A* **2003**, *1016*, 89–98. [CrossRef]
6. Yan, X.; Suzuki, M.; Ohnishi-Kameyama, M.; Sada, Y.; Nakanishi, T.; Nagata, T. Extraction and identification of antioxidants in the roots of yacon (*Smallanthus sonchifolius*). *J. Agric. Food Chem.* **1999**, *47*, 4711–4713. [CrossRef] [PubMed]

7. Lin, F.; Hasegawa, M.; Kodama, O. Purification and identification of antimicrobial sesquiterpene lactones from yacon (*Smallanthus sonchifolius*) leaves. *Biosci. Biotechnol. Biochem.* **2003**, *67*, 2154–2159. [CrossRef] [PubMed]

8. Geyer, M.; Manrique, I.; Degen, L.; Beglinger, C. Effect of yacon (*Smallanthus sonchifolius*) on colonic transit time in healthy volunteers. *Digestion* **2008**, *78*, 30–33. [CrossRef] [PubMed]

9. Valentova, K.; Ulrichova, J. Smallanthus sonchifolius and Lepidium meyenii-prospective andean crops for the prevention of chronic diseases. *Biomed. Pap. Med. Fac. Univ. Palacky Olomouc Czech. Repub.* **2003**, *147*, 119–130. [CrossRef] [PubMed]

10. Valentova, K.; Sersen, F.; Ulrichova, J. Radical scavenging and anti-lipoperoxidative activities of smallanthus sonchifolius leaf extracts. *J. Agric. Food Chem.* **2005**, *53*, 5577–5582. [CrossRef] [PubMed]

11. Scheid, M.M.; Genaro, P.S.; Moreno, Y.M.; Pastore, G.M. Freeze-dried powdered yacon: Effects of fos on serum glucose, lipids and intestinal transit in the elderly. *Eur. J. Nutr.* **2014**, *53*, 1457–1464. [CrossRef] [PubMed]

12. Goto, K.; Fukai, K.; Hikida, J.; Nanjo, F.; Hara, Y. Isolation and structural analysis of oligosaccharides from yacon (*Polymnia sonchifolia*). *Biosci. Biotechnol. Biochem.* **1995**, *59*, 2346–2347. [CrossRef]

13. Hermann, M.; Freire, I.; Pazos, C. Compositional diversity of the yacon storage root. *Ann. Rep. Int. Potato Cent.* **1997**, *98*, 425–432.

14. Niness, K.R. Inulin and oligofructose: What are they? *J. Nutr.* **1999**, *129*, 1402s–1406s. [CrossRef] [PubMed]

15. De Almeida Paula, H.A.; Abranches, M.V.; de Luces Fortes Ferreira, C.L. Yacon (*Smallanthus sonchifolius*): A food with multiple functions. *Crit. Rev. Food Sci. Nutr.* **2015**, *55*, 32–40. [CrossRef] [PubMed]

16. Roberfroid, M.; Gibson, G.R. Prebiotic effects: Metabolic and health benefits. *Br. J. Nutr.* **2010**, *104* (Suppl. 2), S1–S63. [CrossRef] [PubMed]

17. Pedreschi, R.; Campos, D.; Noratto, G.; Chirinos, R.; Cisneros-Zevallos, L. Andean yacon root (*Smallanthus sonchifolius* Poepp. Endl) fructooligosaccharides as a potential novel source of prebiotics. *J. Agric. Food Chem.* **2003**, *51*, 5278–5284. [CrossRef] [PubMed]

18. Kaplan, H.; Hutkins, R.W. Fermentation of fructooligosaccharides by lactic acid bacteria and bifidobacteria. *Appl. Environ. Microbiol.* **2000**, *66*, 2682–2684. [CrossRef] [PubMed]

19. De Souza Lima Sant'Anna, M.; Rodrigues, V.C.; Araujo, T.F.; de Oliveira, T.T.; do Carmo Gouveia Peluzio, M.; de Luces Fortes Ferreira, C.L. Yacon-based product in the modulation of intestinal constipation. *J. Med. Food.* **2015**, *18*, 980–986. [CrossRef] [PubMed]

20. Komura, M.; Fukuta, T.; Genda, T.; Hino, S.; Aoe, S.; Kawagishi, H.; Morita, T. A short-term ingestion of fructo-oligosaccharides increases immunoglobulin A and mucin concentrations in the rat cecum, but the effects are attenuated with the prolonged ingestion. *Biosci. Biotechnol. Biochem.* **2014**, *78*, 1592–1602. [CrossRef] [PubMed]

21. Kato, T.; Fukuda, S.; Fujiwara, A.; Suda, W.; Hattori, M.; Kikuchi, J.; Ohno, H. Multiple omics uncovers host-gut microbial mutualism during prebiotic fructooligosaccharide supplementation. *DNA Res.* **2014**, *21*, 469–480. [CrossRef] [PubMed]

22. Guigoz, Y. Effects of oligosaccharide on the faecal flora and non-specific immune system in elderly people. *Nutr. Res.* **2002**, *22*, 13–25. [CrossRef]

23. Pineiro, M.; Asp, N.G.; Reid, G.; Macfarlane, S.; Morelli, L.; Brunser, O.; Tuohy, K. FAO technical meeting on prebiotics. *J. Clin. Gastroenterol.* **2008**, *42*, S156–S159. [CrossRef] [PubMed]

24. Holloway, L.; Moynihan, S.; Abrams, S.A.; Kent, K.; Hsu, A.R.; Friedlander, A.L. Effects of oligofructose-enriched inulin on intestinal absorption of calcium and magnesium and bone turnover markers in postmenopausal women. *Br. J. Nutr.* **2007**, *97*, 365–372. [CrossRef] [PubMed]

25. Griffin, I.J.; Davila, P.M.; Abrams, S.A. Non-digestible oligosaccharides and calcium absorption in girls with adequate calcium intakes. *Br. J. Nutr.* **2002**, *87*, S187–S191. [CrossRef] [PubMed]

26. Griffin, I.J.; Hicks, P.M.D.; Heaney, R.P.; Abrams, S.A. Enriched chicory inulin increases calcium absorption mainly in girls with lower calcium absorption. *Nutr. Res.* **2003**, *23*, 901–909. [CrossRef]

27. Dornas, W.C.; de Lima, W.G.; Pedrosa, M.L.; Silva, M.E. Health implications of high-fructose intake and current research. *Adv. Nutr.* **2015**, *6*, 729–737. [CrossRef] [PubMed]

28. Takenaka, M.; Yan, X.; Ono, H.; Yoshida, M.; Nagata, T.; Nakanishi, T. Caffeic acid derivatives in the roots of yacon (*Smallanthus sonchifolius*). *J. Agric. Food Chem.* **2003**, *51*, 793–796. [CrossRef] [PubMed]

29. Russo, D.; Valentão, P.; Andrade, P.; Fernandez, E.; Milella, L. Evaluation of antioxidant, antidiabetic and anticholinesterase activities of *Smallanthus sonchifolius* landraces and correlation with their phytochemical profiles. *Int. J. Mol. Sci.* **2015**, *16*, 17696. [CrossRef] [PubMed]

30. Duell, P.B.; Wheaton, D.L.; Shultz, A.; Nguyen, H. Inhibition of LDL oxidation by melatonin requires supraphysiologic concentrations. *Clin. Chem.* **1998**, *44*, 1931–1936. [PubMed]

31. Hondo, M.; Nakano, A.; Okumura, Y.; Yamaki, T. Effects of activated carbon powder treatment on clarification, decolorization, deodorization and fructooligosaccharide content of yacon [*Polymnia sonchifolia*] juice. *J. Jpn. Soc. Food Sci.* **2000**, *47*, 148–154. [CrossRef]

32. Perez-Jimenez, J.; Serrano, J.; Tabernero, M.; Arranz, S.; Diaz-Rubio, M.E.; Garcia-Diz, L.; Goni, I.; Saura-Calixto, F. Bioavailability of phenolic antioxidants associated with dietary fiber: Plasma antioxidant capacity after acute and long-term intake in humans. *Plant Foods Hum. Nutr.* **2009**, *64*, 102–107. [CrossRef] [PubMed]

33. Neves, V.A.; da Silva, M.A. Polyphenol oxidase from yacon roots (*Smallanthus sonchifolius*). *J. Agric. Food Chem.* **2007**, *55*, 2424–2430. [CrossRef] [PubMed]

34. Inoue, A.; Tamogami, S.; Kato, H.; Nakazato, Y.; Akiyama, M.; Kodama, O.; Akatsuka, T.; Hashidoko, Y. Antifungal melampolides from leaf extracts of *Smallanthus sonchifolius*. *Phytochemistry* **1995**, *39*, 845–848. [CrossRef]

35. Lachman, J.; Fernández, E.C.; Orsák, M. Yacon [*Smallanthus sonchifolia* (Poepp. Et Endl.) H. Robinson] chemical composition and use—A review. *Plant Soil Environ.* **2003**, *49*, 283–290. [CrossRef]

36. Sousa, S.; Pinto, J. Antioxidant properties of sterilized yacon (*Smallanthus sonchifolius*) tuber flour. *Food Chem.* **2015**, *188*, 504–509. [CrossRef] [PubMed]

37. Genta, S.; Cabrera, W.; Habib, N.; Pons, J.; Carillo, I.M.; Grau, A.; Sanchez, S. Yacon syrup: Beneficial effects on obesity and insulin resistance in humans. *Clin. Nutr.* **2009**, *28*, 182–187. [CrossRef] [PubMed]

38. Khajehei, F.; Merkt, N.; Claupein, W.; Graeff-Hoenninger, S. Yacon (smallanthus sonchifolius poepp. & endl.) as a novel source of health promoting compounds: Antioxidant activity, phytochemicals and sugar content in flesh, peel, and whole tubers of seven cultivars. *Molecules* **2018**, *23*, 278. [CrossRef]

39. Barcellona, C.S.; Cabrera, W.M.; Honoré, S.M.; Mercado, M.I.; Sánchez, S.S.; Genta, S.B. Safety assessment of aqueous extract from leaf *Smallanthus sonchifolius* and its main active lactone, enhydrin. *J. Ethnopharmacol.* **2012**, *144*, 362–370. [CrossRef] [PubMed]

40. Ma, Z.F.; Lee, Y.Y. Virgin coconut oil and its cardiovascular health benefits. *Nat. Prod. Commun.* **2016**, *11*, 1151–1152.

41. Ma, Z.F.; Zhang, H. Phytochemical constituents, health benefits, and industrial applications of grape seeds: A mini-review. *Antioxidants* **2017**, *6*, 71. [CrossRef] [PubMed]

42. Zhang, H.; Ma, Z.F. Phytochemical and pharmacological properties of *Capparis spinosa* as a medicinal plant. *Nutrients* **2018**, *10*, 116. [CrossRef] [PubMed]

43. Aybar, M.J.; Sanchez Riera, A.N.; Grau, A.; Sanchez, S.S. Hypoglycemic effect of the water extract of *Smallantus sonchifolius* (yacon) leaves in normal and diabetic rats. *J. Ethnopharmacol.* **2001**, *74*, 125–132. [CrossRef]

44. Baroni, S.; Suzuki-Kemmelmeier, F.; Caparroz-Assef, S.M.; Cuman, R.K.N.; Bersani-Amado, C.A. Effect of crude extracts of leaves of *Smallanthus sonchifolius* (yacon) on glycemia in diabetic rats. *Rev. Bras. Cienc. Farm.* **2008**, *44*, 521–530. [CrossRef]

45. Satoh, H.; Audrey Nguyen, M.T.; Kudoh, A.; Watanabe, T. Yacon diet (*Smallanthus sonchifolius*, asteraceae) improves hepatic insulin resistance via reducing Trb3 expression in Zucker fa/fa rats. *Nutr. Diabetes* **2013**, *3*, e70. [CrossRef] [PubMed]

46. Oliveira, P.M.; Coelho, R.P.; Pilar, B.C.; Golke, A.M.; Güllich, A.A.; Piccoli, J.D.C.E.; Manfredini, V. Supplementation with the yacon root extract (*Smallanthus sonchifolius*) improves lipid, glycemic profile and antioxidant parameters in wistar rats hypercholesterolemic. *World J. Pharm. Pharm. Sci.* **2016**, *5*, 2284–2300.

47. Genta, S.B.; Cabrera, W.M.; Grau, A.; Sanchez, S.S. Subchronic 4-month oral toxicity study of dried *Smallanthus sonchifolius* (yacon) roots as a diet supplement in rats. *Food Chem. Toxicol.* **2005**, *43*, 1657–1665. [CrossRef] [PubMed]

48. Habib, N.C.; Honore, S.M.; Genta, S.B.; Sanchez, S.S. Hypolipidemic effect of *Smallanthus sonchifolius* (yacon) roots on diabetic rats: Biochemical approach. *Chem. Biol. Interact.* **2011**, *194*, 31–39. [CrossRef] [PubMed]

49. Oliveira, R.B.; Chagas-Paula, D.A.; Secatto, A.; Gasparoto, T.H.; Faccioli, L.H.; Campanelli, A.P.; Costa, F.B.D. Topical anti-inflammatory activity of yacon leaf extracts. *Rev. Bras. Cienc. Farm.* **2013**, *23*, 497–505. [CrossRef]

50. Schorr, K.; Merfort, I.; Da Costa, F.B. A novel dimeric melampolide and further terpenoids from *Smallanthus sonchifolius* (asteraceae) and the inhibition of the transcription factor NF-κb. *Nat. Prod. Commun.* **2007**, *2*, 367–374.

51. Padla, E.P.; Solis, L.T.; Ragasa, C.Y. Antibacterial and antifungal properties of ent-kaurenoic acid from *Smallanthus sonchifolius. Chin. J. Nat. Med.* **2012**, *10*, 408–414. [CrossRef]

52. Bhandari, A.; Woodhouse, M.; Gupta, S. Colorectal cancer is a leading cause of cancer incidence and mortality among adults younger than 50 years in the USA: A seer-based analysis with comparison to other young-onset cancers. *J. Investig. Med.* **2017**, *65*, 311–315. [CrossRef] [PubMed]

53. Siegel, R.L.; Miller, K.D.; Jemal, A. Cancer statistics, 2016. *CA Cancer J. Clin.* **2016**, *66*, 7–30. [CrossRef] [PubMed]

54. Jasperson, K.W.; Tuohy, T.M.; Neklason, D.W.; Burt, R.W. Hereditary and familial colon cancer. *Gastroenterology* **2010**, *138*, 2044–2058. [CrossRef] [PubMed]

55. Roberts, M.C.; Millikan, R.C.; Galanko, J.A.; Martin, C.; Sandler, R.S. Constipation, laxative use, and colon cancer in a North Carolina population. *Am. J. Gastroenterol.* **2003**, *98*, 857–864. [CrossRef] [PubMed]

56. Chen, H.-L.; Lu, Y.-H.; Lin, J.; Ko, L.-Y. Effects of fructooligosaccharide on bowel function and indicators of nutritional status in constipated elderly men. *Nutr. Res.* **2000**, *20*, 1725–1733. [CrossRef]

57. De Moura, N.A.; Caetano, B.F.; Sivieri, K.; Urbano, L.H.; Cabello, C.; Rodrigues, M.A.; Barbisan, L.F. Protective effects of yacon (*Smallanthus sonchifolius*) intake on experimental colon carcinogenesis. *Food Chem. Toxicol.* **2012**, *50*, 2902–2910. [CrossRef] [PubMed]

58. Rolim, P.M. Development of prebiotic food products and health benefits. *Food Sci. Technol.* **2015**, *35*, 3–10. [CrossRef]

59. Wong, J.M.; de Souza, R.; Kendall, C.W.; Emam, A.; Jenkins, D.J. Colonic health: Fermentation and short chain fatty acids. *J. Clin. Gastroenterol.* **2006**, *40*, 235–243. [CrossRef] [PubMed]

60. Tang, Y.; Chen, Y.; Jiang, H.; Nie, D. The role of short-chain fatty acids in orchestrating two types of programmed cell death in colon cancer. *Autophagy* **2011**, *7*, 235–237. [CrossRef] [PubMed]

61. Vaz-Tostes, M.; Viana, M.L.; Grancieri, M.; Luz, T.C.; Paula, H.; Pedrosa, R.G.; Costa, N.M. Yacon effects in immune response and nutritional status of iron and zinc in preschool children. *Nutrition* **2014**, *30*, 666–672. [CrossRef] [PubMed]

62. Luo, J.; Rizkalla, S.W. Chronic consumption of short-chain fructooligosaccharides by healthy subjects decreased basal hepatic glucose production but had no effect on insulin-stimulated glucose metabolism. *Am. J. Clin. Nutr.* **1996**, *63*, 939–945. [CrossRef] [PubMed]

63. Habib, N.C.; Serra-Barcellona, C.; Honore, S.M.; Genta, S.B.; Sanchez, S.S. Yacon roots (*Smallanthus sonchifolius*) improve oxidative stress in diabetic rats. *Pharm. Biol.* **2015**, *53*, 1183–1193. [CrossRef] [PubMed]

64. Beylot, M. Effects of inulin-type fructans on lipid metabolism in man and in animal models. *Br. J. Nutr.* **2005**, *93*, S163–S168. [CrossRef] [PubMed]

65. Pinto, M.M.; Gonçalez, E.; Rossi, M.H.; Felício, J.D.; Medina, C.S.; Fernandes, M.J.; Simoni, I.C. Activity of the aqueous extract from *Polymnia sonchifolia* leaves on growth and production of aflatoxin B1 by aspergillus flavus. *Braz. J. Microbiol.* **2001**, *32*, 127–129. [CrossRef]

66. Gonçalez, E.; Felicio, J.; Pinto, M.; Rossi, M.; Medina, C.; Fernandes, M.; Simoni, I. Inhibition of aflatoxin production by *Polymnia sonchifolia* and its in vitro cytotoxicity. *Arq. Inst. Biol.* **2003**, *70*, 159–163.

67. Schorr, K.; Da Costa, F.B. A proposal for chemical characterization and quality evaluation of botanical raw materials using glandular trichome microsampling of yacón (*Polymnia sonchifolia*, asteraceae), an andean medicinal plant. *Rev. Bras. Cienc. Farm.* **2003**, *13*, 1–3. [CrossRef]

68. Oliveira, P.M.; Coelho, R.P.; Pilar, B.C.; Golke, A.M.; Güllich, A.A.; Maurer, P.; Piccoli, J.d.C.E.; Schwanz, M.; Manfredini, V. Antioxidative properties of 14-day supplementation with yacon leaf extract in a hypercholesterolemic rat model. *R. Bras. Bioci.* **2018**, *15*, 178–186.

69. Valentova, K.; Moncion, A.; de Waziers, I.; Ulrichova, J. The effect of *Smallanthus sonchifolius* leaf extracts on rat hepatic metabolism. *Cell Biol. Toxicol.* **2004**, *20*, 109–120. [CrossRef] [PubMed]

70. Roberfroid, M.B.; Cumps, J.; Devogelaer, J.P. Dietary chicory inulin increases whole-body bone mineral density in growing male rats. *J. Nutr.* **2002**, *132*, 3599–3602. [CrossRef] [PubMed]

71. Campos, D.; Betalleluz-Pallardel, I.; Chirinos, R.; Aguilar-Galvez, A.; Noratto, G.; Pedreschi, R. Prebiotic effects of yacon (*Smallanthus sonchifolius* Poepp. & Endl), a source of fructooligosaccharides and phenolic compounds with antioxidant activity. *Food Chem.* **2012**, *135*, 1592–1599. [PubMed]
72. Lobo, A.R.; Colli, C.; Alvares, E.P.; Filisetti, T.M. Effects of fructans-containing yacon (*Smallanthus sonchifolius* Poepp and Endl.) flour on caecum mucosal morphometry, calcium and magnesium balance, and bone calcium retention in growing rats. *Br. J. Nutr.* **2007**, *97*, 776–785. [CrossRef] [PubMed]
73. Rodrigues, F.C.; Castro, A.S.; Rodrigues, V.C.; Fernandes, S.A.; Fontes, E.A.; de Oliveira, T.T.; Martino, H.S.; de Luces Fortes Ferreira, C.L. Yacon flour and bifidobacterium longum modulate bone health in rats. *J. Med. Food* **2012**, *15*, 664–670. [CrossRef] [PubMed]
74. Van den Heuvel, E.G.; Muys, T.; van Dokkum, W.; Schaafsma, G. Oligofructose stimulates calcium absorption in adolescents. *Am. J. Clin. Nutr.* **1999**, *69*, 544–548. [CrossRef] [PubMed]
75. Van den Heuvel, E.G.; Schaafsma, G.; Muys, T.; van Dokkum, W. Nondigestible oligosaccharides do not interfere with calcium and nonheme-iron absorption in young, healthy men. *Am. J. Clin. Nutr.* **1998**, *67*, 445–451. [CrossRef] [PubMed]
76. Yun, E.Y.; Kim, H.S. A case of anaphylaxis after the ingestion of yacon. *Allergy Asthma Immunol. Res.* **2010**, *2*, 149–152. [CrossRef] [PubMed]
77. De Oliveira, R.B.; de Paula, D.A.; Rocha, B.A.; Franco, J.J.; Gobbo-Neto, L.; Uyemura, S.A.; dos Santos, W.F.; Da Costa, F.B. Renal toxicity caused by oral use of medicinal plants: The yacon example. *J. Ethnopharmacol.* **2011**, *133*, 434–441. [CrossRef] [PubMed]

MDPI

St. Alban-Anlage 66

4052 Basel

Switzerland

Tel. +41 61 683 77 34

Fax +41 61 302 89 18

www.mdpi.com

Foods Editorial Office

E-mail: foods@mdpi.com

www.mdpi.com/journal/foods